INTRODUCTION TO MATHEMATICAL ANALYSIS

LIBRARY OF MATHEMATICS

edited by

WALTER LEDERMANN

D.Sc., Ph.D., F.R.S.Ed., Professor of
Mathematics, University of Sussex

Linear Equations	P. M. Cohn
Sequences and Series	J. A. Green
Differential Calculus	P. J. Hilton
Elementary Differential Equations and Operators	G. E. H. Reuter
Partial Derivatives	P. J. Hilton
Complex Numbers	W. Ledermann
Principles of Dynamics	M. B. Glauert
Electrical and Mechanical Oscillations	D. S. Jones
Vibrating Systems	R. F. Chisnell
Vibrating Strings	D. R. Bland
Fourier Series	I. N. Sneddon
Solutions of Laplace's Equation	D. R. Bland
Solid Geometry	P. M. Cohn
Numerical Approximation	B. R. Morton
Integral Calculus	W. Ledermann
Sets and Groups	J. A. Green
Differential Geometry	K. L. Wardle
Probability Theory	A. M. Arthurs
Multiple Integrals	W. Ledermann
Fourier and Laplace Transforms	P. D. Robinson
Introduction to Abstract Algebra	C. R. J. Clapham
Functions of a Complex Variable, 2 Vols	D. O. Tall
Linear Programming	Kathleen Trustrum
Sets and Numbers	S. Świerczkoswki

INTRODUCTION TO MATHEMATICAL ANALYSIS

BY

C. R. J. CLAPHAM

Department of Mathematics
University of Aberdeen

ROUTLEDGE & KEGAN PAUL
LONDON, BOSTON AND HENLEY

First published in 1973
by Routledge & Kegan Paul Ltd,
39 Store Street
London WC1E 7DD,
Broadway House, Newtown Road
Henley-on-Thames
Oxon RG9 1EN and
9 Park Street,
Boston, Mass. 02108, U.S.A.
Reprinted in 1979
Printed in Great Britain by
Lowe & Brydone Printers Ltd,
Thetford, Norfolk

ISBN 0 7100 7529 4 (p)

Library of Congress Catalog Card No. 72-95122

Contents

4. Continuous Functions

5. Differentiable Functions

6. The Riemann Integral

Preface

I have tried to provide an introduction, at an elementary level, to some of the important topics in real analysis, without avoiding reference to the central role which the completeness of the real numbers plays throughout. Many elementary textbooks are written on the assumption that an appeal to the completeness axiom is beyond their scope; my aim here has been to give an account of the development from axiomatic beginnings, without gaps, while keeping the treatment reasonably simple. Little previous knowledge is assumed, though it is likely that any reader will have had some experience of calculus.

I hope that the book will give the non-specialist, who may have considerable facility in techniques, an appreciation of the foundations and rigorous framework of the mathematics that he uses in its applications; while, for the intending mathematician, it will be more of a beginner's book in preparation for more advanced study of analysis.

I should finally like to record my thanks to Professor Ledermann for the suggestions and comments that he made after reading the first draft of the text.

University of Aberdeen C. R. J. CLAPHAM

CHAPTER ONE

Axioms for the Real Numbers

1. Introduction

In mathematics we are continually using properties of the real numbers, some of which we consider to be intuitively obvious and others we feel need to be called theorems and furnished with proofs. We shall therefore begin by setting down certain facts about the real numbers for which we shall not demand proofs, and from these **postulates**, or **axioms**, as they are called, we shall deduce all our other results. The first deductions will of course be very elementary, but gradually, as more definitions and notions are introduced, deeper results will be established.

2. Fields

We are familiar with addition, subtraction and multiplication of numbers; these are examples of *operations*. An operation $\circ$ is a rule which associates with two elements a and b, an element denoted by $a \circ b$.

DEFINITION. *A set S is* **closed under the operation** $\circ$ *if, for any two elements a and b in S, there is defined uniquely an element of S denoted by $a \circ b$.*

Example 1. The set **R** of real numbers, as intuitively understood, is closed under addition, subtraction and multiplication. It is not closed under division, for if we take $b = 0$, then $a \div b$ is not defined. The set of *positive* real numbers *is* closed under division, for $a \div b$ is a well-defined positive real number for any positive a and b. The set of positive real numbers is *not* closed under *subtraction*: if $a < b$, then $a - b$ is not a positive number.

Now we take certain postulates about addition and multiplication which hold, intuitively, for the real numbers. But we shall see that there may also be other sets, with two similar operations, that satisfy these same postulates. We give a special name to any set with two operations with these properties:

DEFINITION. A **field** *is a set F which is closed under two operations, called addition and multiplication and denoted in the usual way,* with these properties:*

1. *For all a, b, c in F,* $a+(b+c) = (a+b)+c$. *(Addition is* **associative**.)
2. *For all a, b in F,* $a+b = b+a$. *(Addition is* **commutative**.)
3. *There is an element* 0 *in F such that* $a+0 = a$ *for all a in F. (There is a* **zero** *element.)*
4. *For each a in F, there is an element denoted by* $(-a)$, *such that* $a+(-a) = 0$. *(Each element has a* **negative**.)
5. *For all a, b, c in F,* $a(bc)=(ab)c$. *(Multiplication is associative.)*
6. *For all a, b in F,* $ab = ba$. *(Multiplication is commutative.)*
7. *There is a non-zero element* 1 *in F such that* $a1 = a$ *for all a in F. (There is an* **identity** *element.)*
8. *For each non-zero element a in F, there is an element, denoted by* a^{-1}, *such that* $a^{-1}a = 1$. *(Each non-zero element has an* **inverse**.)

* In this book, we shall keep to the convention of denoting the operation of multiplication by . but normally writing $a.b$ as ab instead.

9. *For all a, b, c in F,* $a(b+c) = ab+ac$. (*Multiplication is* **distributive** *over addition.*)

We shall take it for granted for the moment that the set **R** of real numbers with addition and multiplication as we know them satisfies the necessary properties for a field. But there are also other examples. If we prove that **1** to **9** have certain elementary consequences, these must hold in any field. They will be results we know very well to be true for real numbers.

Example 2. Let **Q** denote the set of all rational numbers, i.e. those real numbers which can be written as fractions m/n, where m and n are integers with $n \neq 0$. (Some real numbers like $\sqrt{2}$ or π are irrational and not contained in **Q**.) Then **Q** with normal addition and multiplication is a field.

Example 3. The set **J** of integers, with the usual addition and multiplication, is not a field because **8** is not satisfied. The integer 2, for example, does not have an inverse because there is no integer x such that $2x = 1$.

Example 4. A set $\mathbf{J}_2$ consisting of two elements denoted by 0 and 1, with addition and multiplication given by $0+0 = 1+1 = 0$, $0+1 = 1+0 = 1$ and $0.0 = 0.1 = 1.0 = 0$, $1.1 = 1$, is a field.

Example 5. The reader who is familiar with complex numbers will be able to appreciate that the set of complex numbers with the standard addition and multiplication is a field.

THEOREM 1. *In any field*

(i) *there is a unique zero element,*

(ii) *there is a unique identity element,*

(iii) *each element has a unique negative,*

(iv) *each non-zero element has a unique inverse.*

Proof. (i) Suppose that 0_1 and 0_2 both have the property stated in **3**. Then $0_2+0_1 = 0_2$ because 0_1 is a zero element, and $0_1+0_2 = 0_1$ because 0_2 is a zero element. But $0_1+0_2 = 0_2+0_1$ by **2**, so $0_1 = 0_2$.

(ii) is proved similarly.

(iii) Suppose that b and c are both the negative of a. Then $a+b = 0$ and $a+c = 0$. So $b+(a+c) = b+0 = b$, by **3**. But also $b+(a+c) = (b+a)+c$ (by **1**) $= (a+b)+c$ (by **2**) $= 0+c$ $= c+0$ (by **2**) $= c$ (by **3**). Therefore $b = c$.

(iv) is proved similarly.

We may now agree to write $a-b$ instead of $a+(-b)$, for subtraction is not an essentially new operation. It is simply a shorthand notation for the sum of a and the negative of b.

THEOREM 2. *In any field F*

(i) $-(-a) = a$, *for all a in F,*

(ii) $a-b = 0$ *if and only if* $a = b$, *for all a and b in F,*

(iii) $a0 = 0$, *for all a in F,*

(iv) $-(ab) = a(-b) = (-a)b$, *for all a and b in F,*

(v) $(-a)(-b) = ab$ *for all a and b in F,*

(vi) $(-1)a = (-a)$ *for all a in F.*

Proof. (i) $(-a)+a = a+(-a)$ (by **2**) $= 0$ (by **4**). Thus a has the property required of the negative of $(-a)$. So $a = -(-a)$.

(ii) If $a = b$, then $a-b = a+(-b) = b+(-b) = 0$. Conversely, if $a-b = 0$, then $a+(-b) = 0$. So $(a+(-b))+b = 0+b = b$, but also $(a+(-b))+b = a+((-b)+b) = a+0 = a$. Hence $a = b$.

(iii) The zero element 0 was given as an element with a special property to do with *addition*. In proving that $a0 = 0$ we see that it has a special *multiplicative* property, so we may expect to use the distributive law that connects addition and multiplication:

$a = a1 = a(0+1) = a0+a1$ (by **9**) $= a0+a$. Therefore $0 = a+(-a) = (a0+a)+(-a) = a0+(a+(-a)) = a0+0 = a0$, i.e. $a0 = 0$.

(iv) $ab+a(-b) = a(b+(-b)) = a0 = 0$ (by (iii)). So $a(-b)$

is the negative of ab, i.e. $a(-b) = -(ab)$. We can show that $(-a)b = -(ab)$ similarly.

(v) $(-a)(-b) = -(a(-b)) = -(-(ab))$ (using (iv) twice) $= ab$ (by (i)).

(vi) $a+(-1)a = 1a+(-1)a = (1+(-1))a = 0a = a0 = 0$. So $(-1)a$ is the negative of a, i.e. $(-1)a = (-a)$.

By **6**, the element $b^{-1}a$ is equal to ab^{-1}, so we may agree to denote either of these by a/b, when $b \neq 0$.

THEOREM 3. *In any field F*

(i) $(a^{-1})^{-1} = a$ *for all* $a \neq 0$ *in* F,

(ii) $a/b = 1$ *if and only if* $a = b$, *for all* a *and* $b(\neq 0)$ *in* F,

(iii) *the equation* $bx = a$, *where* $b \neq 0$, *has the unique solution* $x = a/b$,

(iv) $1/a = a^{-1}$ *for all* $a \neq 0$ *in* F,

(v) *if* $ab = 0$ *then either* $a = 0$ *or* $b = 0$,

(vi) *if* $ab = ac$ *then either* $a = 0$ *or* $b = c$ **(the cancellation law)**,

(vii) $(ab)^{-1} = b^{-1}a^{-1}$, *for all* $a(\neq 0)$ *and* $b(\neq 0)$ *in* F.

Proof. (i) as Theorem 2(i).

(ii) as Theorem 2(ii).

(iii) If $bx = a$, with $b \neq 0$, multiply on the left by b^{-1}. Then $b^{-1}(bx) = b^{-1}a$, so $x = b^{-1}a$. This is therefore the unique solution.

(iv) $1/a = a^{-1}1 = a^{-1}$.

(v) If $ab = 0$ and $a \neq 0$, then a has an inverse a^{-1} and $a^{-1}(ab) = a^{-1}0 = 0$. Hence $(a^{-1}a)b = 0$ and thus $b = 0$.

(vi) If $ab = ac$ and $a \neq 0$, then a has an inverse a^{-1} and $a^{-1}(ab) = a^{-1}(ac)$. Hence $b = c$.

(vii) Since $a \neq 0$ and $b \neq 0$, then $ab \neq 0$ (by (v)), so ab has a unique inverse. But $(b^{-1}a^{-1})(ab) = b^{-1}(a^{-1}a)b = b^{-1}1b = 1$, so $b^{-1}a^{-1}$ is the inverse of ab, as required.

3. Order

The real numbers, of course, have properties that cannot be deduced from the basic postulates for a field. Some of these are to do with the notion of one number being less than another. In other words, there is the idea of a relation $<$ between real numbers, and this relation has various properties that we continually use. We follow our axiomatic approach and set down those properties which we assume to hold, from which all others must be deduced.

DEFINITION. *A field is* **ordered** *if there is a relation $<$ so that for any two elements a and b, either $a < b$ holds or not, with the following properties:*

O1. *If $a < b$, then $a+c < b+c$.*
O2. *If $a < b$ and $0 < c$, then $ac < bc$.*
O3. *For any two elements a and b, one and only one of the following relations holds: $a < b$, $a = b$, $b < a$.*
O4. *If $a < b$ and $b < c$, then $a < c$.*

Example 6. The idea of saying that one real (or rational) number is less than another is familiar and has the properties **O1, O2, O3, O4.** Thus, as commonly understood, **R** and **Q** are ordered fields.

Example 7. The field $\mathbf{J}_2$ (see Example 4) of two elements is not ordered. If it were, apply **O3** to the elements 0 and 1. Since $0 \neq 1$, suppose first that $0 < 1$. Then $0+1 < 1+1$ (by **O1**), i.e. $1 < 0$, contradicting **O3**. The possibility that $1 < 0$ similarly leads to a contradiction.

We may now agree to write $a > b$ if $b < a$, and $a \leq b$ as an abbreviation for '$a < b$ or $a = b$' and $a \geq b$ if $b \leq a$. It is also very reasonable to call certain elements of an ordered field *positive*:

DEFINITION. *In an ordered field, a is* **positive** *if $a > 0$.*

It is easy to see that $a < b$ if and only if $b-a$ is positive.

THEOREM 4. *In an ordered field, the following properties hold:*

P1. *The sum of any two positive elements is positive.*

P2. *The product of any two positive elements is positive.*

P3. *For any element a, one and only one of the following statements is true: a is positive, $a = 0$, $(-a)$ is positive.*

Proof. **P1**. If $a > 0$ and $b > 0$, then $a+b > a+0$ (by **O1**), so $a+b > a$. But $a > 0$, so (by **O4**) $a+b > 0$.

P2. If $a > 0$ and $b > 0$, then $ab > 0$ (by **O2**).

P3. Apply **O3** to the elements a and 0. Either $a > 0$ or $a = 0$ or $0 > a$. The third possibility, on adding $(-a)$ to both sides, becomes $(-a) > 0$.

The following properties of the *negative* elements can also be established:

DEFINITION. *In an ordered field, a is* **negative** *if $(-a)$ is positive.*

THEOREM 5. *In an ordered field*

(i) *if a is positive, $(-a)$ is negative,*

(ii) *a is negative if and only if $a < 0$,*

(iii) *the sum of two negative elements is negative,*

(iv) *the product of two negative numbers is positive,*

(v) *the product of a negative element and a positive element is negative.*

The proof is left as an exercise.

Example 8. Many important properties follow from the two previous theorems:

(i) *Any non-zero element squared is positive.* Suppose that $a \neq 0$. Then either a is positive or negative, by **P3**. By definition, $a^2 = a.a$, so if a is positive, a^2 is positive by **P2**. On the other hand, if a is negative, then a^2 is again positive, by Theorem 5 (iv).

(ii) *The identity element is positive.* The identity element 1 is equal to $1.1 = 1^2$. Thus 1 is a square and so is positive.

(iii) *If* $0 < a$, *then* $0 < \frac{1}{2}a < a$. (Here $\frac{1}{2}$ denotes the inverse of 2, where $2 = 1+1$.) Since $0 < 1$, **O1** gives $0+1 < 1+1$, so $1 < 2$. So 2 is also positive. Now $\frac{1}{2}.2 = 1$, and, with 1 and 2 both positive, so also is $\frac{1}{2}$. From $0 < 1 < 2$, we may now deduce $0 < \frac{1}{2} < 1$, using **O2**. Finally, if a is any positive element, then $0 < \frac{1}{2}a < a$, using **O2** again.

(iv) We have shown that, given any positive element, there is a smaller positive element; consequently, *there is no smallest positive element.*

Example 9. The field **C** of complex numbers is not ordered. In any ordered field, 1 is positive, so -1 is negative, but in **C**, $-1 = i^2$ and so, being the square of a non-zero element, would be positive if **C** were ordered.

DEFINITION. *If a is an element of an ordered field, the* **absolute value** *of a, denoted by $|a|$, is a itself if $a \geq 0$ and $-a$ if $a < 0$.*

Thus $|a|$ is positive except when $a = 0$.

THEOREM 6.

(i) $|ab| = |a|\,|b|$,

(ii) $|a+b| \leq |a|+|b|$,

(iii) $|a-b| \geq ||a|-|b||$.

Proof. (i) We consider the different possibilities. If a or b is zero, then the left-hand side $|ab|$ and the right-hand side $|a|\,|b|$ are both zero and so equal. If a and b are both positive, both sides equal ab. If a and b are both negative, the right-hand side is equal to $(-a)(-b)$ and the left-hand side equals ab, since ab is positive; so the two sides are equal by Theorem 2(v). If a, say, is positive and b is negative, the right-hand side equals $a(-b)$ and the left-hand side is equal to $-(ab)$, since ab is negative; so the two sides are equal by Theorem 2(iv).

(ii) Certainly $a \leq |a|$ and $b \leq |b|$, so $a+b \leq |a|+|b|$. Also $-a \leq |a|$ and $-b \leq |b|$, so $-(a+b) = (-a)+(-b) \leq |a|+|b|$. But $|a+b|$ is either $a+b$ or $-(a+b)$, so $|a+b| \leq |a|+|b|$.

(iii) Let $x = a-b$. Then $|a| = |b+x| \leq |b|+|x|$, by (ii), so $|a-b| = |x| \geq |a|-|b|$. Similarly $|b-a| \geq |b|-|a|$. So $|a-b|$, which equals $|b-a|$, is greater than or equal to both $|a|-|b|$ and $|b|-|a|$. Hence $|a-b| \geq ||a|-|b||$.

4. Completeness

Since the familiar idea of the rational numbers and the real numbers both satisfy the postulates for an ordered field (and there are other examples, too), it is clear that there are not yet enough postulates to characterize the real numbers uniquely. Rather surprisingly, the essence of the real numbers lies in just one more property they have, their *completeness*, and this can be captured in just one further definition:

DEFINITION. *An ordered field F is* **complete** *if the following holds:*

10. *Whenever A and B are non-empty subsets of F such that*
(i) *every element of F is either in A or B but not both,*
(ii) *every element of A is less than every element of B,*
then there is an element ξ of F such that ξ is greater than or equal to every element of A and ξ is less than or equal to every element of B. (This is **Dedekind's axiom.**)

Notice that it is not specified whether the element ξ is in A or B, but it must be in one or the other.

Example 10. It is possible to prove, using only elementary facts about integers that there is no *rational* number satisfying $x^2 = 2$. That is to say, the real number $\sqrt{2}$ is irrational. Hence if we divide the set **Q** of rational numbers into the two subsets

$$A = \{x | x \leq 0, \text{ or } x > 0 \text{ and } x^2 < 2\},$$
$$B = \{x | x > 0 \text{ and } x^2 > 2\},$$

then every rational number is in A or B but not both, and every number in A is less than every number in B. But there is no 'dividing' number ξ as required in the above definition. Therefore **Q** is not a complete ordered field.

We shall not be able to *prove*, however, that **R** is a complete ordered field, for the reader will notice that we have not rigorously defined the real numbers, or indeed the rational numbers or integers. When we have referred to them, we have assumed that the reader has understood what we have been talking about. We have now reached the point, however, when we can be quite precise: it will be an *assumption* from now on that *the real numbers form a complete ordered field.* We cannot prove that such a structure exists, except that our intuitive idea of the real numbers appears to do so. It is possible, though, to prove that there is, at most, essentially only one such object. What this means needs to be clarified and we shall not do that here. It amounts to this: given two complete ordered fields, the elements of one can be made to correspond with the elements of the other in such a way that any property involving the elements of one is shared by the corresponding elements of the other. This is not necessary, though, for our present purpose. We have written down the properties that we shall assume the real numbers to have, and from these we can develop the whole subject that lies before us.

R contains two special elements 0 and 1, and it is natural to denote $1+1$ by 2, $2+1$ by 3, $3+1$ by 4, and so on, and the negatives of these in the usual way. It is easy to show that in any ordered field these elements are all different, and these particular elements of **R** will be called the **integers**. With this rather loose definition, we shall moreover suppose that it is clear that if a set S has the property that 1 is in S and when k is in S then $k+1$ is in S, then S contains all the positive integers. Indeed, we could define the set of positive integers as the intersection of all sets with this property. We can then deduce the **principle of induction**:

THEOREM 7. *Let there be associated with each positive integer n,* *a proposition $P(n)$ which is either true or false. If $P(1)$ is true and, for all k, $P(k)$ implies $P(k+1)$, then $P(n)$ is true for all positive integers n.*

Proof. The set of those n for which $P(n)$ is true satisfies the conditions for S above, and so it contains all the positive integers.

Finally, if a and b are integers, with $b \neq 0$, then a/b, which is the unique solution of the equation $bx = a$, is called a **rational number.** We can then establish familiar results about rational numbers such as these: $a/b = c/d$ if and only if $ad = bc$, $a/b + c/d = (ad+bc)/bd$ and $(a/b)(c/d) = (ac/bd)$.

5. Upper bound

DEFINITION. *If S is a non-empty subset of* **R**, *and b is a real number greater than or equal to every element of S, then b is an* **upper bound** *of S. If S has an upper bound, then S is* **bounded above.**

If S is a non-empty subset of **R** *and c is a real number less than or equal to every element of S, then c is a* **lower bound** *of S. If S has a lower bound, then S is* **bounded below.**

If S is bounded above and bounded below, S is **bounded.**

Notice that if b is an upper bound for S, then any greater number than b is also an upper bound.

If S is bounded above, we very often want to choose a *least* upper bound; we shall define what is meant by this and then discover that we need to *prove* that it exists:

DEFINITION. *The real number b is a* **supremum** (*or* **least upper bound**) *of S if*

(i) *b is an upper bound of S,*

(ii) *there is no upper bound of S less than b.*

We write $b = \sup S$.

The real number c is an **infimum** (*or* **greatest lower bound)** *of S if*

(i) *c is a lower bound of S,*

(ii) *there is no lower bound of S greater than c.*

We write $c = \inf S$.

Example 11. If S is finite, then the largest element of S is $\sup S$. So in this case $\sup S$ belongs to S. But if S is, say, the set of all real numbers less than 5, then $\sup S = 5$ and $\sup S$ does not belong to S.

We need to prove that when S is bounded above, there *is* a supremum; for the argument that says 'of all the upper bounds, take the smallest' might be just as impossible an instruction to obey as 'of all the positive numbers, take the smallest' (see Example 8(iv)). In fact, the existence of a least upper bound for every non-empty set bounded above is equivalent to Dedekind's axiom, and so is by no means trivial:

THEOREM 8. **(The completeness theorem).**

(i) *If S is a non-empty subset of* **R** *bounded above, then S has a supremum.*

(ii) *If S is a non-empty subset of* **R** *bounded below, then S has an infimum.*

Proof. (i) Let B be the set of upper bounds of S, and let A consist of all numbers not in B. Certainly B is not empty because S has an upper bound, and A is not empty because if x is an element of S, $x-1$ is certainly in A. Now suppose that a is any element of A and b any element of B. Then a is not greater than or equal to every element of S, so for some s in S $a < s$. But b is an upper bound and so $s < b$. Thus $a < b$. Hence the conditions of Dedekind's axiom are satisfied. Let ξ

be the number whose existence is guaranteed by that axiom. Suppose, if possible, that ξ is in A. Then there is an s in S such that $\xi < s$. But then $\xi < \frac{1}{2}(\xi+s) < s$. So $\frac{1}{2}(\xi+s)$ is an element of A bigger than ξ. This is a contradiction.

Thus ξ is in B. So ξ is an upper bound. But ξ is less than or equal to every element of B, so there is no upper bound of S less than ξ. Hence ξ is a supremum.

(ii) This may be proved in a similar way to (i), or instead it may be deduced from (i) without resorting to Dedekind's axiom again (see Exercise 11).

An alternative course of action is to take, as the axioms for a complete ordered field, the following: **1** to **9**, **O1**, **O2**, **O3**, **O4** and

10′. *Every non-empty subset of F which is bounded above has a supremum.*

10′ replaces **10**, and can be called, in this approach, the completeness axiom. It is then necessary to show that the property described in **10** follows from these axioms, and this result is then called Dedekind's theorem (see Exercise 12).

6. The Archimedean property

THEOREM 9. **(Archimedes' theorem).** *If $x > 0$, then for any real number y, there is an integer n such that $nx > y$.*

Proof. Suppose that there is no integer n such that $nx > y$. Then the set $S = \{nx \mid n \text{ is an integer}\}$ is bounded above by y. Let $b = \sup S$. Since $b-x$ is less than b, $b-x$ is not an upper bound of S, so for some integer k, say, $b-x < kx$. But then $b < (k+1)x$, so there is an element of S bigger than b. This contradiction shows that the initial assumption is false, and the theorem is proved.

THEOREM 10.

(i) *For any real number y, there is an integer n such that $n > y$.*

(ii) *If $x > 0$, then there is a positive integer n such that $1/n < x$.*

Proof. (i) is the particular case of Theorem 9 with $x = 1$, and (ii) is that with $y = 1$.

THEOREM 11. *Given any real number x, there is a unique integer n such that $x-1 < n \leq x$.*

Proof. Let h be an integer less than $x-1$. If $x-1 > 0$ we can take $h = 0$; if $x-1 \leq 0$ then $-(x-1) \geq 0$ and there is an integer $h' > -(x-1)$, by Theorem 10(i). Then $-h' < x-1$, so take $h = -h'$.

Now suppose there are no integers n satisfying $x-1 < n \leq x$. Let $P(m)$ be the proposition '$h+m \leq x$'. $P(1)$ says: $h+1 \leq x$, which is true since $h < x-1$. Now suppose $P(k)$ is true: $h+k \leq x$. Then $h+k \leq x-1$ for it is impossible, we are supposing, that $x-1 < h+k \leq x$. Hence $h+k+1 \leq x$, i.e. $P(k+1)$ is true. Thus, by the principle of induction, $h+m \leq x$ for all positive integers m. But there is an integer $N > x$ (by Theorem 10(i)) and so $h+(N-h) > x$. This gives a contradiction because $N > h$, and consequently $N-h$ is a positive integer and $P(N-h)$ is false. Thus the supposition that there were no integers in the given interval was false.

The proof will be complete if the reader shows (Exercise 14) that there cannot be more than one integer in the interval.

DEFINITION. *The unique number n given by Theorem* 11 *is called the* **integer part** *of x, and is denoted by* $[x]$.

Example 12. $[1] = 1$, $[\frac{1}{2}] = 0$, $[\sqrt{2}] = 1$. Take care with negative numbers, however: $[-1] = -1$, $[-\frac{1}{2}] = -1$, $[-\sqrt{2}] = -2$.

THEOREM 12. *Given two real numbers x and y, with $x < y$, there is a rational number between them.*

Proof. We want to find integers p and q such that $x < p/q < y$, i.e. such that $qx < p < qy$. So choose q such that $q > 1/(y-x)$ (by Theorem 10(i)). Hence $qy-qx > 1$, i.e. $qy > qx+1$. Using the integer part just defined, let $p = [qx+1]$, so that $qx < p \leq qx+1$. Then $qx < p < qy$, and p/q lies between x and y as required.

Example 13. The way in which any real number can be expressed as a (possibly infinite) decimal is something we tend to take for granted. Let us see how, with our approach, this is obtained. For simplicity, we shall confine ourselves to finding the decimal expansion for a real number x between 0 and 1. We shall use the integers between 0 and 9 inclusive as **digits**: let $D = \{0, 1, 2, 3, 4, 5, 6, 7, 8, 9\}$.

If x is a real number such that $0 \leq x < 1$, then $0 \leq 10x < 10$. Let $x_1 = [10x]$, the integer part of $10x$. Certainly x_1 belongs to D. Then $0 \leq 10x - x_1 < 1$, so $0 \leq 10(10x-x_1) < 10$. Let $x_2 = [100(x-x_1/10)]$, so that x_2 belongs to D. Continuing this process by defining, at each stage,

$$x_r = [10^r(x-x_1/10-x_2/10^2- \ldots - x_{r-1}/10^{r-1})],$$

it is not difficult to show that each x_r belongs to D. Then $\cdot x_1x_2 \ldots x_r \ldots$ is called the **decimal expansion** of x.

EXERCISES

1. Show that the set of real numbers of the form $a+b\sqrt{2}$, where a and b are rational, with the natural definition of addition and multiplication, is a field.

2. Prove that the following rules hold in any field, justifying each step by a postulate, a result established in the text or an earlier exercise:

   (i) If $a+b = a+c$, then $b = c$.

   (ii) If $a+b = a$, then $b = 0$.

(iii) $-(a+b) = (-a)+(-b)$.

(iv) $-(a-b) = b-a$.

(v) If $ab = a$, then $b = 1$.

(vi) $(ab^{-1})^{-1} = ba^{-1}$.

(vii) $a^{-1}(bc^{-1}) = b(ac)^{-1}$.

3. Show that, in any field, the only elements satisfying $x^2 = x$ are 0 and 1.

4. Prove that the following rules hold in any ordered field:

(i) $a-b < a-c$ if and only if $b > c$.

(ii) If $a < b$ and $c < 0$, then $ac > bc$.

(iii) If $a < b$, then $-a > -b$.

(iv) If $a > 0$ and $b > 0$, $a > b$ if and only if $a^2 > b^2$.

5. Provide the proofs for Theorem 5(i) to (v).

6. Suppose that F is a field in which certain elements are called positive such that **P1**, **P2** and **P3** hold. Define the relation $<$ by $a < b$ if and only if $b-a$ is positive. Show that $<$ satisfies **O1**, **O2**, **O3** and **O4**, and hence that F is ordered.

7. Prove that, in an ordered field, if $r > 0$, then $|x-a| < r$ if and only if $a-r < x < a+r$.

8. Find, if they exist, the supremum and infimum of the following subsets of **R**:

(i) $\{x \mid x > 0 \text{ and } 2 < x^2 < 4\}$,

(ii) $\{x \mid x = n/(n+1) \text{ for some positive integer } n\}$,

(iii) $\{x \mid x < 0 \text{ and } x^2 = n \text{ for some positive integer } n\}$,

(iv) $\{x \mid x \text{ is rational and } x^2 < 2\}$,

(v) $\{x \mid x^2+3x+2 = 0\}$.

9. Show that a non-empty subset S of **R** is bounded if and only if there is a number M such that $|x| \leq M$ for all x in S.

10. Show that if x is rational and y is irrational, then $x+y$ and xy are irrational.

11. If S is a non-empty subset of **R** bounded below and A is the set of all lower bounds of S, A is a non-empty set bounded above, which therefore has a supremum ξ, by Theorem 8(i). Show that ξ is an infimum of S.

12. On the assumption that **R** satisfies axioms **1** to **9**, **O1**, **O2**, **O3** **O4**

and the completeness axiom: **10′**, every non-empty subset which is bounded above has a supremum, prove that **R** has the property **10** described as Dedekind's axiom.

13. Use the method of Theorem 9 to prove that, for any real number y, there is an integer n such that $2^n > y$.

14. Prove that there is at most one integer n satisfying $x-1 < n \leq x$ (see end of proof of Theorem 11).

CHAPTER TWO

Sequences

---

7. Limit of a sequence

The notion of a limit, fundamental to analysis, is most easily introduced in terms of sequences of real numbers. An infinite **sequence** is a succession of real numbers $a_1, a_2, a_3, \ldots$ in a definite order, with a **term** a_n corresponding to each positive integer n. This sequence will be denoted by (a_n).

We now want to consider what the terms a_n are like when n is large, and describe the different situations that can arise. It may be, for example, that the terms approach a certain fixed number, and we want to say that a number A is the *limit* of the sequence if *eventually* all the terms in the sequence are *as close as we please* to A. More precisely:

DEFINITION. *The sequence* (a_n) *has the limit* A *if, given any positive number* ϵ *(however small), there is an integer* N *(which depends on* ϵ*) such that all the terms in the sequence after the N-th lie between* $A-\epsilon$ *and* $A+\epsilon$*. We write* $\lim (a_n) = A$.

If we want to make use of notation that is available, this can be abbreviated to 'given $\epsilon > 0$, there is an integer N such that $|a_n - A| < \epsilon$ for all $n > N$'.

If $\lim (a_n) = A$, we sometimes write '$a_n \to A$' and say 'a_n tends to A'.

Example 14. A first, almost trivial, example of a limit is that of a constant sequence, where $a_n = k$ for all n. In this case $\lim (a_n) = k$, for given any positive number ϵ we may take $N = 1$, because *all* the terms of the sequence are between $k-\epsilon$ and $k+\epsilon$.

THEOREM 13. *If* $a_n = 1/n$, *then* $\lim (a_n) = 0$.

Proof. Immediately we write this in terms of the definition, we see that it is a consequence of Archimedes' theorem:

$\lim (a_n) = 0$ if, given $\epsilon > 0$, there is an integer N such that $|a_n - 0| < \epsilon$, i.e. such that $1/n < \epsilon$, for all $n > N$.

Theorem 10(ii) tells us that there is certainly an integer N such that $1/N < \epsilon$. Moreover, if $n > N$, then by simple division $1/n < 1/N$, so $1/n < \epsilon$. Thus all the terms in the sequence after the Nth lie between $-\epsilon$ and ϵ, as required.

We may, instead, often use Theorem 10(i) to prove from first principles that sequences have certain limits.

Example 15. If $a_n = 1/(n+1)$, then $\lim (a_n) = 0$. Given $\epsilon > 0$, we must find N such that $1/(n+1) < \epsilon$ for all $n > N$. This inequality holds if $n+1 > 1/\epsilon$, i.e. if $n > (1/\epsilon)-1$. So if N is an integer greater than $(1/\epsilon)-1$, e.g. $N = [1/\epsilon]$, then when $n > N$, it is certainly the case that $-\epsilon < 1/(n+1) < \epsilon$ as required.

Example 16. If $a_n = (3n+2)/n$, then $\lim (a_n) = 3$. Given $\epsilon > 0$, we want to ensure that $|(3n+2)/n - 3| < \epsilon$, i.e. that $|2/n| < \epsilon$. This is so if $n > 2/\epsilon$. We may therefore take, as the required N, any integer greater than $2/\epsilon$, e.g. $N = [2/\epsilon]+1$. The existence of such an N for any given ϵ implies that the limit of the given sequence is indeed 3.

DEFINITION. *A sequence* (a_n) *is* **bounded above** *if the set* $\{a_n, \text{ for all } n\}$, *consisting of all numbers appearing as terms in the sequence, is bounded above, i.e. if there is a number M such that* $a_n \leq M$ *for all n. There is a corresponding definition of* **bounded below**, *and a sequence which is bounded above and below is* **bounded.**

THEOREM 14. *If a sequence* (a_n) *has a limit, then it is bounded.*

Proof. Suppose that $a_n \to A$; let us prove that (a_n) is bounded above. We shall argue that certainly, from some point on, $a_n < A+1$, and up to that point, there are only a finite number of terms, so these are necessarily bounded above. In detail, because $a_n \to A$, then (taking a value of ϵ equal to 1) there is an integer N such that for all $n > N$, $A-1 < a_n < A+1$. So if we put M equal to the greatest of the following $N+1$ numbers: $a_1, a_2, \ldots, a_N, A+1$, we see that for all n, $a_n \leq M$. The reader should show similarly that the sequence is bounded below.

Now, rather than using first principles to consider any example, we can establish a few simple rules to find the limits of given sequences from other limits already known.

THEOREM 15. *If* (a_n) *and* (b_n) *are sequences with limits* A *and* B *respectively, then*

(i) (a_n+b_n) *has the limit* $A+B$,
(ii) (a_nb_n) *has the limit* AB,
(iii) *if* $b_n \neq 0$ *for all* n *and* $B \neq 0$, (a_n/b_n) *has the limit* A/B.

Proof. (i) To show that (a_n+b_n) has the limit $A+B$, we must suppose that we are given $\epsilon > 0$, and show that an N can be found such that $(A+B)-\epsilon < a_n+b_n < (A+B)+\epsilon$, for all $n > N$. We know that $a_n \to A$, so take the number $\frac{1}{2}\epsilon$ and let N_1 be the corresponding number such that $|a_n-A| < \frac{1}{2}\epsilon$, for all $n > N_1$. Similarly, since $b_n \to B$, let N_2 be the number such that $|b_n-B| < \frac{1}{2}\epsilon$, for all $n > N_2$. Consequently, if N is the greater of N_1 and N_2, when $n > N$, both $|a_n-A| < \frac{1}{2}\epsilon$ and $|b_n-B| < \frac{1}{2}\epsilon$. So $|(a_n+b_n)-(A+B)| = |(a_n-A)+(b_n-B)| \leq |a_n-A|+|b_n-B|$ (by Theorem 6(ii)) $< \frac{1}{2}\epsilon+\frac{1}{2}\epsilon = \epsilon$. The required N has thus been found.

(ii) First of all, notice that $a_nb_n-AB = (a_n-A)b_n+A(b_n-B)$, so $|a_nb_n-AB| \leq |a_n-A|\,|b_n|+|A|\,|b_n-B|$. Given $\epsilon > 0$, we want to make the left-hand side of this inequality less than ϵ, so we try to make each term on the right-hand side less than $\frac{1}{2}\epsilon$.

Since (b_n) has a limit, it is bounded, by Theorem 14, so $|b_n| \leq M$, say, for all n. So considering the number $\epsilon/2M$, take N_1 such that for all $n > N_1$, $|a_n - A| < \epsilon/2M$. This is possible since $a_n \to A$. Since $b_n \to B$, take N_2 such that for all $n > N_2$, $|b_n - B| < \epsilon/2|A|$. Then, if N is the greater of N_1 and N_2, when $n > N$, $|a_n b_n - AB| \leq |a_n - A|M + |A|\ |b_n - B| < (\epsilon/2M)M + |A|(\epsilon/2|A|) = \frac{1}{2}\epsilon + \frac{1}{2}\epsilon = \epsilon$, as required.

(iii) It is enough to show that $(1/b_n)$ has the limit $1/B$, for the required result follows by applying (ii) to the two sequences (a_n) and $(1/b_n)$. We shall suppose that $B > 0$, and leave the reader to make the necessary changes for the case when $B < 0$. We need to consider

$$\left|\frac{1}{b_n} - \frac{1}{B}\right|, \text{ which is equal to } \frac{|b_n - B|}{|b_n|B}.$$

Since $b_n \to B$, there is an N_1 such that when $n > N_1$, b_n lies between $B - \frac{1}{2}B$ and $B + \frac{1}{2}B$, i.e. such that $\frac{1}{2}B < b_n < \frac{3}{2}B$. The first of these inequalities gives $b_n B > \frac{1}{2}B^2$. But also there is an N_2 such that when $n > N_2$, $|b_n - B| < \frac{1}{2}B^2\epsilon$. Consequently, if N is the greater of N_1 and N_2, when $n > N$, we have, as required,

$$\left|\frac{1}{b_n} - \frac{1}{B}\right| < \frac{\frac{1}{2}B^2\epsilon}{\frac{1}{2}B^2} = \epsilon.$$

Example 17. Notice the particular case of Theorem 15(ii), when one of the sequences is constant, say $b_n = k$, and hence $B = k$ (Example 14). *If (a_n) has the limit A, then the sequence (ka_n) has the limit kA.* For example, $1/n \to 0$, so $2/n \to 2.0 = 0$.

Example 18. If $a_n = (3n+2)/n$ (see Example 16), then without resorting to first principles we have $a_n = 3 + (2/n)$ and, as we have just seen, $2/n \to 0$, so, by Theorem 15(i), $3 + (2/n) \to 3 + 0 = 3$.

Example 19. Show that $(3n+2)/(n^2+5) \to 0$. We use here each part of Theorem 15. Dividing numerator and denominator by n^2 gives

$$\frac{(3/n)+(2/n^2)}{1+(5/n^2)}.$$

Now $1/n \to 0$, so certainly $1/n^2 = (1/n)(1/n) \to 0.0$ (by (ii)) $= 0$. So $3/n \to 0$, $2/n^2 \to 0$ and $5/n^2 \to 0$. Hence $(3/n)+(2/n^2) \to 0+0 = 0$ and $1+(5/n^2) \to 1+0 = 1$, using (i). Finally, by (iii),

$$\frac{(3/n)+(2/n^2)}{1+(5/n^2)} \to \frac{0}{1} = 0.$$

8. Sequences without limits

We have spoken so far only of sequences that have limits; but it is obviously very easy to write down sequences that have no limit.

Example 20. (n^2), the sequence 1, 4, 9, 16, . . . , has no limit.

Example 21. $(-n-3)$, the sequence $-4, -5, -6, -7, \ldots$, has no limit.

Example 22. $((-1)^n n/(n+1))$, the sequence $-\frac{1}{2}, \frac{2}{3}, -\frac{3}{4}, \frac{4}{5}, \ldots$, has no limit.

Example 23. The sequence 1, 2, 1, 4, 1, 8, 1, 16, . . . has no limit.

It is useful to classify these according to the kind of behaviour they show. In Example 20, for instance, the terms of the sequence become arbitrarily large in the following sense:

DEFINITION. *We say that $a_n \to \infty$ ('a_n tends to infinity') if, given any number K (however large), there is an integer N (which depends on K) such that all the terms in the sequence after the N-th are greater than K.*

Though we say a_n tends to infinity in this case, we still say that (a_n) has no limit, for on no account is ∞ to be treated as a real number.

Example 24. For the sequence 1, 4, 9, 16, . . . , the term $a_n = n^2 \to \infty$, because $n^2 > K$ if $n > \sqrt{K}$; so if we put $N = [\sqrt{K}]+1$, all the terms in the sequence after the Nth are greater than K.

DEFINITION. *We say that $a_n \to -\infty$ ('a_n tends to minus infinity') if, given any number K, there is an integer N such that all the terms in the sequence after the N-th are less than K.*

Example 25. For the sequence $-4, -5, -6, \ldots$, the term $a_n = -n-3 \to -\infty$, for $-n-3 < K$ if $n > -K-3$; so $N = [-K-3]+1$ may be taken as the required N.

The remaining sequences, those that have no limit and do not tend to ∞ or $-\infty$, can be divided into two kinds:

DEFINITION. *If a sequence (a_n) does not have a limit, and a_n does not tend to ∞ or $-\infty$, then (a_n) is* **oscillating.** *If the sequence is bounded, it* **oscillates finitely,** *and if not it* **oscillates infinitely.**

Example 26. The sequence $-\frac{1}{2}, \frac{2}{3}, -\frac{3}{4}, \frac{4}{5}, \ldots$ oscillates finitely. The sequence is bounded above by 1 and below by -1, but, since the even terms tend to 1 and the odd terms tend to -1, the sequence clearly has no limit.

Example 27. The sequence 1, 2, 1, 4, 1, 8, ... oscillates infinitely. Given K, we can find an N such that $a_n > K$, since $a_{2r} = 2^r$, which is greater than K for a suitable r (see Exercise 13, Chapter 1). Thus (a_n) is not bounded above. But a_n does not tend to ∞ because, given K, we cannot find an N such that *all* the terms after the Nth are greater than K.

Example 28. The sequence $(-1)^n/n$, which is $-1, \frac{1}{2}, -\frac{1}{3}, \frac{1}{4}, \ldots$, may at first sight appear to oscillate finitely. But, in fact, it can be seen that this sequence has the limit 0.

9. Monotone sequences

DEFINITION. *The sequence (a_n) is* **increasing** *if $a_n \leq a_{n+1}$ for all n. Similarly, a* **decreasing** *sequence (a_n) is one for which $a_n \geq a_{n+1}$ for all n. A sequence which is either increasing or decreasing is* **monotone.**

It is perhaps intuitively obvious that an increasing sequence of real numbers which is bounded above, or a decreasing sequence bounded below, must have a limit. But if we are to give a rigorous proof of this fact, we notice that the result is stating that a real number with a certain property exists, and a proof of existence is very likely to depend on Dedekind's axiom or its equivalent. In fact, the proof is a very straightforward consequence of the completeness theorem, for the limit is none other than the supremum of the sequence:

THEOREM 16. **(The fundamental theorem on monotone sequences).** *An increasing sequence which is bounded above has a limit.*

Proof. If (a_n) is bounded above, then the set $\{a_n$, for all $n\}$ has supremum A, say. To prove that A is the limit of the sequence, suppose that a positive number ϵ is given. Then certainly one of the terms of the sequence, a_N, say, is greater than $A-\epsilon$, for otherwise $A-\epsilon$ is an upper bound smaller than A. So $a_N > A-\epsilon$. But as the sequence is increasing, all subsequent terms are greater than or equal to a_N, so $a_n > A-\epsilon$ for all $n > N$. But, for all n, $a_n \leq A$, since A is an upper bound, so we certainly have, for $n > N$, $A-\epsilon < a_n < A+\epsilon$. The existence of such an N, for any given ϵ, is exactly what is required in order that A be the limit of the sequence.

In the same way, we can prove that a decreasing sequence which is bounded below has a limit.

Example 29. If x is a fixed number with $-1 < x < 1$, then $x^n \to 0$. Suppose that $0 < x < 1$. Let $a_n = x^n$, so that the sequence (a_n) we are considering is $x, x^2, x^3, \ldots$. Now $a_n - a_{n+1} = x^n - x^{n+1} = x^n(1-x) > 0$, since $x > 0$ and $1-x > 0$. So $a_n > a_{n+1}$ and (a_n) is a decreasing sequence. But all the terms are positive, so it is bounded below by zero, and consequently (a_n) has a limit A, say.

Now let $b_n = x^{n+1}$. Then (b_n) is the same sequence as (a_n) without the first term, so (see Exercise 2(i)) has the same limit A. But $b_n = xa_n$, so by Example 17, $b_n \to xA$. Therefore $A = xA$, i.e. $(1-x)A = 0$. But $x \neq 1$, so $A = 0$, as required.

Suppose now that $-1 < x < 0$. Let $a_n = x^n$. Then $|a_n| = |x^n| = |x|^n$, using Theorem 6(i), and we know that $|x|^n \to 0$ since $0 < |x| < 1$. Thus $|a_n| \to 0$, and it follows that $a_n \to 0$ as required, by Exercise 3.

EXERCISES

1. Prove that a sequence cannot have two different limits.

2. Prove that

(i) if $a_n \to A$ and $b_n = a_{n+1}$, then $b_n \to A$

(ii) if $a_n \to A$ and $b_n = a_{2n}$, then $b_n \to A$

(iii) if $a_n \to A$ and $b_n \to A$ and $c_{2n} = a_n$, $c_{2n+1} = b_n$, then $c_n \to A$.

3. Prove that if $a_n \to A$ then $|a_n| \to |A|$. [Hint: use Theorem 6(iii).] Give an example to show that, in general, $|a_n| \to |A|$ does not imply that $a_n \to A$. Show, however, that if $|a_n| \to 0$, then $a_n \to 0$.

4. Prove that if $a_n \leq b_n \leq c_n$, for all n, and $a_n \to A$ and $c_n \to A$, then $b_n \to A$.

5. Prove that if $a_n \to 0$ and (b_n) is bounded, and if $c_n = a_n b_n$, then $c_n \to 0$.

6. Generalize Exercise 13, Chapter 1, to prove that if $x > 1$, then for any real number y, there is an integer n such that $x^n > y$. Deduce that, if $x > 1$, $x^n \to \infty$. Describe the behaviour of (x^n) when

(i) $x = 1$

(ii) $x = -1$

(iii) $x < -1$.

7. If $a_n = r^{1/n}$, where $r > 1$, show that (a_n) is decreasing and bounded below by 1, and hence has a limit $A \geq 1$. By considering the sequence (b_n), where $b_n = a_{2n}^2$, find A.

8. Construct infinitely oscillating sequences (a_n) and (b_n) such that if $c_n = a_n + b_n$, then

(i) (c_n) has a limit

(ii) (c_n) oscillates finitely

(iii) (c_n) oscillates infinitely.

9. If (a_n) and (b_n) oscillate finitely and $c_n = a_n + b_n$, what behaviour is possible for (c_n)? Give an example of each.

10. If $a_n \neq 0$ for all n and $b_n = 1/a_n$, prove that if $a_n \to \infty$ then $b_n \to 0$. Give an example to show that, in general, $a_n \to 0$ does not imply that $b_n \to \infty$. Show, however, that if $a_n > 0$ for all n and $b_n = 1/a_n$, then $a_n \to 0$ does imply that $b_n \to \infty$.

11. If $a_n = \sqrt{(n^2+n)} - n$, show that $0 < a_n < \frac{1}{2}$, and prove from first principles that $a_n \to \frac{1}{2}$.

12. Prove that if $a_n > 0$ and $a_{n+1}/a_n < k$ for all n, where $k < 1$, then $a_n \to 0$. Hence show that $a_n \to 0$, where

(i) $a_n = (n+1)/2^{2n+1}$

(ii) $a_n = n^2/2^n$.

13. Give an example of a sequence (a_n) where $a_n > 0$ and $a_{n+1}/a_n < 1$ for all n, and $a_n \to l$ where $l \neq 0$.

CHAPTER THREE

Series

10. Infinite series

It follows from the associative law for addition that, if $a_1, \ldots, a_N$ are a finite number of elements, the sum $a_1+a_2+\ldots+a_N$ gives the same element whichever way brackets are inserted, so this expression has an unambiguous meaning. We now want to give a meaning, if possible, to an expression like $a_1+a_2+a_3+\ldots$, where (a_n) is an infinite sequence. A convenient notation for a finite sum $a_1+a_2+\ldots+a_N$ is $\sum_{n=1}^{N} a_n$, and we shall denote the infinite series $a_1+a_2+a_3+\ldots$ by $\sum_{n=1}^{\infty} a_n$ or just $\sum a_n$. In some cases it is possible to find a number which can reasonably be called the *sum* of the series. In other cases, we can write down the series but cannot define its sum in any sensible way.

11. Convergence

As well as the sequence (a_n) of terms, we can associate, with a series $\sum a_n$, another sequence which is really the more important. Let $s_n = a_1+a_2+\ldots+a_n$, the sum of the first n terms of the series. Then s_n is the nth **partial sum**, and it is the behaviour of the sequence (s_n) of partial sums which is critical. If (s_n) has a limit S, then we say the series $\sum a_n$ is **convergent** and its **sum** is S. If (s_n) does not have a limit, the series is **divergent.**

Example 30. The series $\sum_{n=1}^{\infty} 1/n(n+1)$ is convergent and its sum is 1. Here $a_n = 1/n(n+1) = (1/n)-(1/(n+1))$. So $s_n = a_1+a_2+\ldots+a_n = (1/1-1/2)+(1/2-1/3)+\ldots+((1/n)-(1/(n+1))) = 1-(1/(n+1))$. (This can be established rigorously by induction.) Hence $s_n \to 1$ and $\sum a_n$ is convergent with sum 1.

Example 31. Show that if x is a fixed number with $-1 < x < 1$ then $\sum_{n=0}^{\infty} x^n$ is convergent, and find its sum. (This is a **geometric series.**) (Notice that it is sometimes convenient to have the first term given by $n = 0$ instead of $n = 1$.)

The nth partial sum $s_n = 1+x+\ldots+x^{n-1}$, so $xs_n = x+x^2+\ldots+x^n$. Therefore $s_n - xs_n = 1-x^n$. Since $x \neq 1$, $s_n = (1-x^n)/(1-x) = 1/(1-x) - x^n/(1-x)$. Now $x^n \to 0$ (see Example 29) and so $s_n \to 1/(1-x)$. Therefore $\sum_{n=0}^{\infty} x^n$ is convergent if $-1 < x < 1$, and its sum is $1/(1-x)$.

Example 32. The series $\sum_{n=1}^{\infty} (1/n)$ is divergent. (This is the **harmonic series.**) We consider just *some* of the partial sums:

$$s_2 = 1 + \frac{1}{2} = \frac{3}{2},$$

$$s_4 = s_2 + \frac{1}{3} + \frac{1}{4} > s_2 + \left(\frac{1}{4} + \frac{1}{4}\right) = s_2 + \frac{1}{2} = \frac{4}{2},$$

$$s_8 = s_4 + \frac{1}{5} + \frac{1}{6} + \frac{1}{7} + \frac{1}{8} > s_4 + \left(\frac{1}{8} + \frac{1}{8} + \frac{1}{8} + \frac{1}{8}\right) = s_4 + \frac{1}{2} > \frac{5}{2},$$

$$s_{16} = s_8 + \frac{1}{9} + \frac{1}{10} + \ldots + \frac{1}{16} > s_8 + \left(\frac{1}{16} + \frac{1}{16} + \ldots + \frac{1}{16}\right)$$

$$= s_8 + \frac{1}{2} > \frac{6}{2}.$$

We find that, for $k > 1$, $s_{2^k} > \frac{1}{2}(k+2)$. It is left to the reader to prove this by induction. This shows that, by taking k large enough, we can make s_{2^k} bigger than any given number. Thus the sequence (s_{2^k}) is unbounded. If (s_n) had a limit, then by Theorem 14 it would be bounded, so there would be a number M such that $|s_n| \leq M$ for all n. But this would imply that

$s_{2^k}| \leq M$, for all k, which is not so. Therefore (s_n) does not have a limit and $\sum(1/n)$ is divergent.

12. Tests

THEOREM 17. *If $\sum a_n$ is convergent, then $a_n \to 0$.*

Proof. If $\sum a_n$ is convergent, then (s_n) has a limit S, and hence (s_{n-1}) also has S as a limit. (We think of (s_{n-1}) as a sequence whose first term is s_2 and are using here the result of Exercise 2(i), Chapter 2.) Therefore $a_n = s_n - s_{n-1} \to S - S = 0$.

This theorem shows that if a_n does not tend to zero, then $\sum a_n$ is not convergent. But if a_n does tend to zero, it does not necessarily mean that $\sum a_n$ is convergent. The harmonic series shows that that conclusion is false.

However, there are methods of proving that a series is convergent without necessarily finding what its sum is, and we shall obtain some of these *tests*. The two that follow are the most basic and can be used to prove the convergence of many series; they rely on the fundamental theorem on monotonic sequences.

TEST 1. **(The comparison test)**. *Let $\sum a_n$ and $\sum b_n$ be two series with non-negative terms (i.e. $a_n \geq 0$, $b_n \geq 0$ for all n) such that $a_n \leq b_n$ for all n.*

(i) *If $\sum b_n$ is convergent, then $\sum a_n$ is convergent.*
(ii) *If $\sum a_n$ is divergent, then $\sum b_n$ is divergent.*

Proof. (i) Let $s_n = a_1 + a_2 + \ldots + a_n$, $t_n = b_1 + b_2 + \ldots + b_n$. Because $a_n \geq 0$ and $b_n \geq 0$, (s_n) and (t_n) are increasing sequences, and $t_n \to T$, say, because $\sum b_n$ is convergent. Moreover, T is sup $\{t_n\}$, so $t_n \leq T$ for all n. But $a_n \leq b_n$ for all n, so $s_n \leq t_n$ for all n, and hence $s_n \leq T$ for all n. Thus (s_n), an increasing sequence bounded above, must tend to a limit S, say, and $\sum a_n$ is convergent. Incidentally, $S \leq T$.

(ii) By (i), $\sum b_n$ convergent would imply $\sum a_n$ convergent.

Example 33. $\sum(1/\sqrt{n})$. Because $\sqrt{n} \leq n$, $(1/n) \leq (1/\sqrt{n})$. We have shown that $\sum(1/n)$ is divergent, so $\sum(1/\sqrt{n})$ is divergent by Test 1(ii).

Example 34. $\sum(1/n^2)$. We prove the convergent of this, by comparison with $\sum 1/n(n+1)$ (see Example 30). But it is not a straightforward comparison. We must consider instead $\sum 1/(n+1)^2$, which converges by comparison (i.e. using Test 1(i)) with $\sum 1/n(n+1)$, because $1/(n+1)^2 \leq 1/n(n+1)$ for all n. We then notice that $\sum_{n=1}^{\infty} 1/n^2$ is the same series as $\sum_{n=1}^{\infty} 1/(n+1)^2$ with an additional term at the beginning, so it converges too.

Example 35. $\sum(1/n^3)$. For all n, $n^3 \geq n^2$, so $1/n^3 \leq 1/n^2$. Thus, by comparison, $\sum 1/n^3$ is convergent.

TEST 2. **(The ratio test).** *Let $\sum a_n$ be a series with positive terms, such that a_{n+1}/a_n tends to a limit l. Then*

(i) *if $l > 1$, $\sum a_n$ is divergent,*

(ii) *if $l < 1$, $\sum a_n$ is convergent.*

Proof. (i) If $l > 1$, then $a_{n+1}/a_n > 1$ eventually, i.e. for all $n \geq N$, say. So, from this point on, $a_{n+1} > a_n$. Since the terms are all positive, it is impossible that $a_n \to 0$, so $\sum a_n$ diverges, by Theorem 17.

(ii) If $l < 1$, let k be the number $\frac{1}{2}(1+l)$. Then $l < k < 1$. Since $a_{n+1}/a_n \to l$, we can be sure that $a_{n+1}/a_n < k$ eventually, i.e. for all $n \geq N$, say. So, from this point on, $a_{n+1} < ka_n$. Therefore $a_{N+1} < ka_N$, $a_{N+2} < k^2 a_N, \ldots, a_{N+p} < k^p a_N$. Now $k < 1$, so the geometric series $\sum_{p=1}^{\infty} k^p$ is convergent. So a constant multiple of this, $\sum a_N k^p$, is convergent (by Exercise 1(iii)). By Test 1, $\sum_{p=1}^{\infty} a_{N+p}$ is convergent. This is the series $a_{N+1} + a_{N+2} + \ldots$, and adding just a finite number of terms at the beginning gives the full series, which is therefore convergent (by Exercise 3).

Example 36. $\sum(1/n!)$ is convergent. For the ratio $a_{n+1}/a_n = (n!)/(n+1)! = 1/(n+1) \to 0$. Thus the series $1+1/1!+1/2!+1/3!+\ldots$ is convergent, and one way to define the number e, an important constant in mathematics, is to define it as the sum of this series.

Example 37. If $0 < x < 1$, $\sum x^n/n^2$ is convergent. Since $x > 0$, the series has positive terms and $a_{n+1}/a_n = (x^{n+1}/(n+1)^2)/(x^n/n^2) = xn^2/(n+1)^2 \to x$. So if $x < 1$, the series is convergent.

Example 38. The series $1-\frac{1}{2}+\frac{1}{3}-\frac{1}{4}+\ldots$, with signs $+$ and $-$ alternately, is called the **alternating harmonic series.** Its convergence can be established as follows:

First consider the sequence (s_{2n-1}) of odd partial sums, $s_1, s_3, s_5, \ldots$. Because $s_{2n+1}-s_{2n-1} = a_{2n}+a_{2n+1} = (-1/2n)+1/(2n+1) < 0$, this sequence is decreasing. Consider also the sequence (s_{2n}) of even partial sums, $s_2, s_4, s_6, \ldots$. Now, $s_{2n}-s_{2n-2} = a_{2n-1}+a_{2n} = 1/(2n-1)-1/2n > 0$. So this sequence is increasing.

Next notice that $s_{2n+1}-s_{2n} = 1/(2n+1) > 0$, so $s_{2n+1} > s_{2n} \geq s_2(=\frac{1}{2})$. Thus the decreasing sequence of odd partial sums is bounded below, and so s_{2n-1} tends to a limit S, say. Similarly, the increasing sequence of even partial sums is bounded above by $s_1(=1)$, so $s_{2n} \to S'$, say. Now $s_{2n-1} = s_{2n}-a_{2n}$, so $S = \lim s_{2n-1} = \lim s_{2n} - \lim a_{2n} = S'-0 = S'$. So this common value is the limit of the combined sequence (s_n) of odd and even partial sums (see Exercise 2(iii), Chapter 2), and the series is convergent.

The reader should show as an exercise that any alternating series, $\sum a_n$, where the terms are alternately positive and negative, is convergent provided that (i) $a_n \to 0$ (ii) $|a_n| \geq |a_{n+1}|$ for all n. (These conditions for convergence are known as **Leibniz's test.**)

13. Absolute convergence

THEOREM 18. *If the series* $\sum|a_n|$ *is convergent, then* $\sum a_n$ *is convergent.*

Proof. Write $p_n = \frac{1}{2}(|a_n|+a_n)$ and $q_n = \frac{1}{2}(|a_n|-a_n)$. If $a_n > 0$, then $p_n = a_n$ and $q_n = 0$. On the other hand, if $a_n \leq 0$, $p_n = 0$ and $q_n = -a_n$. So $0 \leq p_n \leq |a_n|$ and $0 \leq q_n \leq |a_n|$ for all n. By the comparison test, if $\sum|a_n|$ is convergent, so are $\sum p_n$ and $\sum q_n$. But $a_n = p_n - q_n$, so $\sum a_n$ is also convergent (see Exercise 1(ii)).

The converse of this theorem is not true: the alternating harmonic series $1-\frac{1}{2}+\frac{1}{3}-\frac{1}{4}+\ldots$ is an example of a series $\sum a_n$ which is convergent, with $\sum|a_n|$ not convergent. This leads us to distinguish between the two situations that can arise:

DEFINITION. *The series* $\sum a_n$ *is* **absolutely convergent** *if* $\sum|a_n|$ *is convergent. If, on the other hand,* $\sum a_n$ *is convergent but* $\sum|a_n|$ *is not, then* $\sum a_n$ *is* **conditionally convergent.**

14. Power series

A series of the form $c_0+c_1x+c_2x^2+\ldots$, where the terms involve ascending powers of a variable x, is a **power series.**

Example 39. $\sum_{n=0}^{\infty} x^n$, the series $1+x+x^2+\ldots$, is a power series and is convergent if and only if $|x| < 1$.

Example 40. $\sum_{n=0}^{\infty} x^n/n!$, the power series $1+x+x^2/2!+x^3/3!+\ldots$, is convergent for all values of x.

Example 41. $\sum n!x^n$, the power series $1+x+2!x^2+3!x^3+\ldots$, is convergent only when $x = 0$.

We shall see that a power series must behave in one of the three ways illustrated by these examples. First, however, a preliminary result:

THEOREM 19. *If the power series* $\sum c_ny^n$ *is convergent, and* $|x| < |y|$, *then* $\sum c_nx^n$ *is absolutely convergent.*

Proof. Since $\sum c_ny^n$ converges, the terms certainly tend to zero, so we can find a number M such that $|c_ny^n| < M$, for all n. Then $|c_nx^n| < M|x/y|^n$. Since $|x/y| < 1$, the geometric series $\sum|x/y|^n$ is convergent and so $\sum M|x/y|^n$ is also. The series $\sum|c_nx^n|$ is convergent by the comparison test.

THEOREM 20. *For any given power series* $\sum c_n x^n$, *there are just three possibilities:*

(i) *It is absolutely convergent for all values of* x.

(ii) *There is a real number* R *such that the series is absolutely convergent for all* x *with* $|x| < R$, *and divergent for all* x *with* $|x| > R$.

(iii) *It is convergent only if* $x = 0$.

Proof. Let S be the set of x for which the series is convergent. Every power series is convergent for $x = 0$, so S contains the number 0. If S contains all the positive real numbers, then, using Theorem 19, it contains all the negative real numbers, too, and we have case (i).

So we may suppose that there is some positive real number x_0 which is not in S. Then S does not contain any number larger than x_0, again by Theorem 19. Thus S is bounded above by x_0. Let R be sup S. If $R = 0$, we have case (iii). So we may suppose $R > 0$. We prove this has the two properties required in (ii).

First, suppose that $|x| < R$. Choose a number x_1 such that $|x| < x_1 < R$. The series $\sum c_n x_1^n$ is convergent, for otherwise x_1 is not in S and no larger number than x_1 is in S, making x_1 an upper bound for S and contradicting the choice of R. By Theorem 19, $\sum c_n x^n$ is absolutely convergent.

Second, suppose that $|x| > R$. Choose a number x_2 such that $R < x_2 < |x|$. If $\sum c_n x^n$ were convergent, then, by Theorem 19, $\sum c_n x_2^n$ would be convergent and x_2 would belong to S, contradicting the choice of R. Therefore $\sum c_n x^n$ is divergent.

DEFINITION. *If case* (ii) *of Theorem* 20 *arises, the number* R *is the* **radius of convergence** *of the series. This terminology is extended to case* (i) *when we say that the radius of convergence is infinite and to case* (ii) *when we say that the radius of convergence is zero. The open interval* $(-R, R)$ *is the* **interval of convergence.**

Notice that in case (ii), the theorem does not say whether the series is convergent or divergent for the two values $x = R$ and $x = -R$. In a great many cases, the radius of convergence can be found using the ratio test, but the behaviour at the two **end-points** requires special investigation.

Example 42. $\sum_{n=1}^{\infty} (-1)^{n+1}x^n/n$. Consider the modified series $\sum_{n=1}^{\infty} |(-1)^{n+1}x^n/n|$ and use the ratio test:

$$\frac{a_{n+1}}{a_n} = \left|\frac{(-1)^{n+2}x^{n+1}}{n+1} \cdot \frac{n}{(-1)^{n+1}x^n}\right| = \frac{n|x|}{(n+1)} \to |x|.$$

If $|x| < 1$, then the modified series is convergent and the original series is absolutely convergent. If $|x| > 1$, the modified series is divergent. So when $|x| > 1$ the original series is not absolutely convergent. Thus the radius of convergence of the series is 1. The end-points are $x = 1$ and $x = -1$. When $x = 1$, we get the series $1-\frac{1}{2}+\frac{1}{3}-\frac{1}{4}+\ldots$, which is convergent (Example 38), and when $x = -1$, we get the series $\sum(-1/n)$, which is $-1-\frac{1}{2}-\frac{1}{3}-\frac{1}{4}-\ldots$. This is divergent (Example 32).

EXERCISES

1. Prove that if $\sum a_n$ is convergent and $\sum b_n$ is convergent, then so is $\sum c_n$, where

(i) $c_n = a_n+b_n$

(ii) $c_n = a_n-b_n$

(iii) $c_n = ka_n$, k a constant.

2. Let $c_n = a_n+b_n$, for all n. Prove that if $\sum a_n$ is convergent and $\sum b_n$ is divergent, then $\sum c_n$ is divergent.

3. Prove that if, for some fixed p, $b_n = a_{n+p}$, then $\sum a_n$ is convergent if and only if $\sum b_n$ is convergent.

4. By showing that $1/(3n+2) > 1/5n$ and $1/(3n^2+2) < 1/3n^2$, prove that $\sum 1/(3n+2)$ is divergent and $\sum 1/(3n^2+2)$ is convergent.

5. Establish whether the following series are convergent or divergent:

$$\sum \frac{4}{n^4},\quad \sum \frac{n+2}{n^2},\quad \sum \frac{1}{n3^n},\quad \sum \frac{2}{2^n+1},\quad \sum \frac{n-3}{n^4},$$

$$\sum \frac{3^n}{2^n},\quad \sum \frac{n^4}{2^n},\quad \sum \frac{n\sqrt{n}}{n^2+1},\quad \sum \frac{n!}{1.3.5\ldots(2n-1)}.$$

6. Prove that the series $\sum(x^n/n!)$ is absolutely convergent for all x, and deduce that, for all x, $x^n/n! \to 0$.

7. Prove that the series $\sum(n^k/x^n)$ is convergent, where $k > 0$ and $x > 1$. Deduce that $n^k/x^n \to 0$.

8. Establish whether the following series are convergent or divergent:

$$\sum(-1)^n 2^n/n!,\qquad \sum(-2)^n,\qquad \sum(-1)^n/\sqrt{n},\qquad \sum(-1)^n(n+1)/n.$$

Find out whether those that are convergent are absolutely convergent or conditionally convergent.

9. Divergent series can be divided into those that diverge properly to infinity, diverge properly to minus infinity, oscillate finitely or oscillate infinitely, according to the behaviour of the sequence (s_n) of partial sums. Give an example of each kind.

10. Find the radius of convergence of the following power series and ascertain the behaviour at the end-points, where appropriate:

$$\sum \frac{(-2x)^n}{3^{2n}},\quad \sum n!\left(\frac{x}{2}\right)^n,\quad \sum \frac{(2x)^n}{n!},\quad \sum \frac{(2n^2+1)x^n}{3^n(n+5)^3}.$$

CHAPTER FOUR

Continuous Functions

15. Limit of a function

If f is a function which associates with a real number x a corresponding real number y, we write $y = f(x)$ and call the set of real numbers x for which the function does this the **domain** of the function.

The behaviour of a function as x gets large (and positive) is akin to the behaviour of sequences. For example $f(x)$ may have a limit:

DEFINITION. *We say $f(x) \to l$ as $x \to \infty$ if, given a positive number ϵ (however small), there is a number X (which depends on ϵ) such that, for all $x > X$, $f(x)$ lies between $l-\epsilon$ and $l+\epsilon$.*

Or else the function may behave like a sequence where $a_n \to \infty$:

DEFINITION. *We say $f(x) \to \infty$ as $x \to \infty$ if, given a positive number K (however large), there is a number X (which depends on K) such that, for all $x > X$, $f(x) > K$.*

But the behaviour of the function may require investigation not only as x gets large, but also as x gets nearer to a certain value a. We want to say that $f(x) \to l$ as $x \to a$ if we can make $f(x)$ as close as we please to l by confining x to a sufficiently small 'neighbourhood' of a:

DEFINITION. *$f(x) \to l$ as $x \to a$ if, given any positive number (however small), there is a positive number δ (which depends on ϵ) such that for all x, except possibly for a itself, lying between $a-\delta$ and $a+\delta$, we have $f(x)$ between $l-\epsilon$ and $l+\epsilon$. We write '$\lim_{x\to a} f(x) = l$'.*

Notice that a itself may not be in the domain of f, i.e. $f(a)$ may not be defined. But there must be a neighbourhood of a, the whole of which (except possibly a itself) is in the domain of f.

Example 43. If $f(x) = (2x^2-x-6)/(x-2)$, $(x \neq 2)$, then the domain of f is the whole of the set of real numbers except 2. But we can show that $f(x) \to 7$ as $x \to 2$. The proof from first principles is as follows. Given $\epsilon > 0$, we want to make $|(2x^2-x-6)/(x-2) - 7|$ less than ϵ, that is to say $|(2x^2-8x+8)/(x-2)| < \epsilon$. So we want $|(2(x-2)^2)/(x-2)| < \epsilon$, which is so if $|x-2| < \frac{1}{2}\epsilon$. Thus we may take $\delta = \frac{1}{2}\epsilon$, because then, if $0 < |x-2| < \delta$, we have ensured that $|f(x)-7| < \epsilon$ as required.

Of course, in this example, we could also look at, say, $\lim_{x\to 3} f(x)$, but we would find that it was nothing but $f(3)$, i.e. 9. However, if we define $f(2) = 0$, say, so that the domain of f is now the whole of **R**, we see that '$\lim_{x\to 2} f(x) \neq f(2)$.

16. Continuity

DEFINITION. *The function f is* **continuous at** *a if $f(x) \to f(a)$ as $x \to a$.*

This property ensures that the function is 'well-behaved' at a particular point: intuitively speaking, it means that the function does not suddenly jump there. We can extend this very simply to cover a whole interval. An **open** interval (a, b) is the set of x such that $a < x < b$; a **closed** interval $[a, b]$ is the set of x such that $a \leq x \leq b$.

DEFINITION. *f is* **continuous in an open interval** *if it is continuous at each point of the interval.*

To deal with a closed interval we must introduce the idea of left and right limits which will be needed at the end-points:

DEFINITION. *We say that* $f(x) \to l$ *as* $x \to a$ **from the right** *if, given* $\epsilon > 0$, *there is a number* δ *such that for all* x *strictly between* a *and* $a+\delta$, *we have* $f(x)$ *between* $l-\epsilon$ *and* $l+\epsilon$. *We write* '$\lim_{x \to a+} f(x) = l$'.

We say that $f(x) \to l$ *as* $x \to a$ **from the left** *if, given* $\epsilon > 0$, *there is a number* δ *such that for all* x *strictly between* $a-\delta$ *and* a, *we have* $f(x)$ *between* $l-\epsilon$ *and* $l+\epsilon$. *We write* '$\lim_{x \to a-} f(x) = l$'.

DEFINITION. f *is* **continuous in the closed interval** $[a, b]$, *where* $a < b$, *if it is continuous in the open interval* (a, b) *and if* $\lim_{x \to a+} f(x) = f(a)$ *and* $\lim_{x \to b-} f(x) = f(b)$.

Example 44. From first principles, we can easily establish that if $f(x) = k$, for all x, then f is continuous at any point a, and hence also in any interval, open or closed.

If now f is the function defined by $f(x) = x$ for all x, then f is continuous (at any point or in any interval). Given $\epsilon > 0$, we can take $\delta = \epsilon$ because, with that choice, if x lies between $a-\delta$ and $a+\delta$, $f(x)$ lies between $f(a)-\epsilon$ and $f(a)+\epsilon$. So $f(x) \to f(a)$ as $x \to a$.

Having obtained some continuous functions we can build up many more using the following theorems:

(i) *The sum of two continuous functions is continuous.*
(ii) *The product of two continuous functions is continuous.*
(iii) *The quotient of two continuous functions is continuous for any value of* x *where the denominator is not zero.*

The reader should show that these results can be proved in a similar way to the results for sequences in Theorem 15 (see also Exercise 5).

We can now prove that if n is a positive integer, the function f defined by $f(x) = x^n$ is continuous, by using induction and applying (ii) to the functions x^{n-1} and x. Then by multiplying by constants and using (i) to build up sums, we conclude that *any polynomial is continuous at any point.*

From (iii), it follows that a quotient of two polynomials is continuous at all points where the denominator is not zero.

For example, if $f(x) = (2x^2-x-6)/(x-2)$ (see Example 43), f is continuous for all $x \neq 2$.

Example 45. The function defined by $f(x) = (2x^2-x-6)/(x-2)$, $(x \neq 2)$, and $f(2) = 0$ (see Example 43) is discontinuous at $x = 2$.

Example 46. The function $[x]$ is discontinuous for all integer values of x; if n is any integer, $\lim_{x \to n+} [x] = n$, but $\lim_{x \to n-} [x] = n-1$, and it is clear that a function f is continuous at a if and only if $\lim_{x \to a+} f(x) = f(a)$ and $\lim_{x \to a-} f(x) = f(a)$.

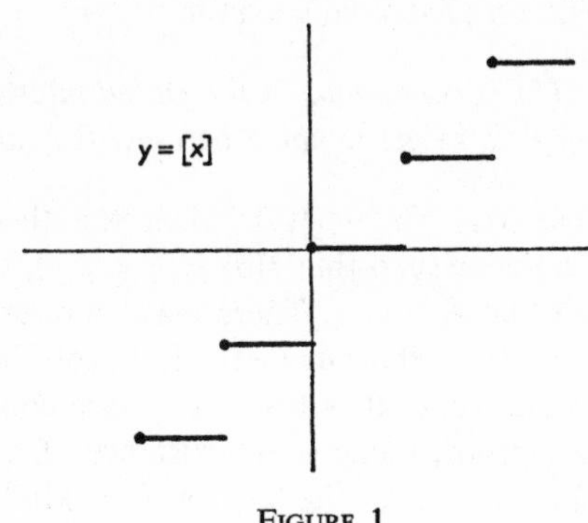

FIGURE 1

THEOREM 21. *Suppose that f is continuous at a and $f(a) = b$, and g is continuous at b. If h is the composite $g \circ f$ of f and g, defined by $h(x) = g(f(x))$ wherever this is meaningful, then h is continuous at a.*

Proof. Because g is continuous at b, given $\epsilon > 0$, there is a positive number γ, say, such that $|g(y)-g(b)| < \epsilon$ whenever $|y-b| < \gamma$. But, because f is continuous at a, to this number γ, there is a corresponding positive number δ such that $|f(x)-f(a)| < \gamma$, whenever $|x-a| < \delta$. Combining these two, we see that if $|x-a| < \delta$, then $|f(x)-f(a)| < \gamma$, i.e. $|f(x)-b| < \gamma$ and so $|g(f(x))-g(b)| < \epsilon$, i.e. $|g(f(x))-g(f(a))| < \epsilon$, or $|h(x)-h(a)| < \epsilon$. Consequently, h is continuous at a.

17. The intermediate value property

We show now that functions which are continuous in an interval have very 'strong' properties. If we think of a continuous function as one whose graph is a continuous line, then the properties are 'obvious', but the proofs are not elementary, relying as they do on Dedekind's axiom.

THEOREM 22. *If f is continuous in the closed interval $[a, b]$ and $f(a) \neq f(b)$, then f takes every value between $f(a)$ and $f(b)$.*

Proof. Suppose that $f(a) < f(b)$. Then the theorem states that if η is any number such that $f(a) < \eta < f(b)$, then there is a number ξ such that $f(\xi) = \eta$. There may, of course, be more than one such number ξ, but our method of proof has the effect of giving the greatest one. It is because we are concerned here with an existence proof, namely the existence of a solution of the equation $f(x) = \eta$, that the use of Dedekind's axiom is likely to be required.

Let S be the set of x such that $f(x) < \eta$. Then S is not empty because it contains a, and since S is bounded above by b, S has a supremum ξ, say. We prove first that ξ is strictly between a and b.

As f is 'continuous on the right' at a, choosing $\epsilon = \eta - f(a)$, there is a corresponding value δ such that if $a < x < a+\delta$, $f(x)$ lies between $f(a)-\epsilon$ and $f(a)+\epsilon$. So within that interval $(a, a+\delta)$, $f(x) < f(a)+\epsilon = \eta$. Therefore $\xi \geq a+\delta$, i.e. ξ is strictly greater than a. Similarly, ξ is strictly less than b.

Now we prove that $f(\xi) = \eta$:

If $f(\xi) < \eta$, then, choosing now $\epsilon = \eta - f(\xi)$, the continuity of f at ξ implies that there is a number δ such that if x lies between $\xi-\delta$ and $\xi+\delta$, $f(x)$ lies between $f(\xi)-\epsilon$ and $f(\xi)+\epsilon$. So there is a value $\xi+\frac{1}{2}\delta$, for instance, which is greater than ξ, such that $f(\xi+\frac{1}{2}\delta) < f(\xi)+\epsilon = \eta$. This contradicts the choice of ξ as sup S.

On the other hand, if $f(\xi) > \eta$, then with $\epsilon = f(\xi)-\eta$ now, the continuity at ξ implies that there is a δ such that when x is between $\xi-\delta$ and $\xi+\delta$, $f(x)$ lies between $f(\xi)-\epsilon$ and $f(\xi)+\epsilon$. So, whenever $\xi-\delta < x \leq \xi$, $f(x)$ is greater than $f(\xi)-\epsilon = \eta$. None of the points of this interval therefore lies in S, so sup $S \leq \xi-\delta$, a contradiction.

Thus we must have $f(\xi) = \eta$.

The reader should see that a similar proof may be given if $f(a) > f(b)$.

18. Bounds of a continuous function

THEOREM 23. *If f is continuous in a closed interval, then f is bounded.*

Proof. We shall show that f is bounded above; a proof that f is bounded below is similar. So let S be the set of all s less than or equal to b, such that $f(x)$ is bounded above in the closed interval $[a, s]$. What we want to show is that S contains b.

S is certainly not empty, because it contains a. Moreover, all the members of S are less than or equal to b, so let $\xi =$ sup S.

We first show that $\xi > a$. Because f is continuous on the right at a, taking $\epsilon = 1$, there is an interval $(a, a+\delta)$ throughout which $f(x) < f(a)+1$. Thus $f(x)$ is bounded above, by $f(a)+1$, throughout this interval and so $a+\frac{1}{2}\delta$, for example, is definitely in S. So $\xi \geq a+\frac{1}{2}\delta > a$.

Now suppose that $\xi < b$. Then the continuity of f at ξ implies that, with $\epsilon = 1$ again, there is within $[a, b]$ an interval $(\xi-\delta, \xi+\delta)$ throughout with $f(x) < f(\xi)+1$. (We are using the fact that $\xi > a$.) But because $\xi =$ sup S, there must be an element s of S in $(\xi-\delta, \xi)$. Then f is bounded above in $[a, s]$, so suppose that $f(x) \leq M$ for all x in $[a, s]$. Then, throughout the closed interval $[a, \xi+\frac{1}{2}\delta]$, we have that $f(x)$ is less than or equal to the greater of M and $f(\xi)+1$. So $\xi+\frac{1}{2}\delta$ is in S, a contradiction.

The only possibility is that $\xi = b$, and f is bounded above over the whole interval $[a, b]$.

Having shown that a continuous function (over a *closed* interval) is bounded, we now show that such a function 'attains its bounds'.

THEOREM 24. *Let f be continuous in a closed interval* $[a, b]$ *and let S be the set of values $f(x)$ for x in* $[a, b]$. *If* $M = \sup S$, *then there is a ξ in* $[a, b]$ *such that* $f(\xi) = M$ *(and similarly for* $m = \inf S$*).*

Proof. Suppose that there is no ξ in $[a, b]$ such that $f(\xi) = M$. Then $M-f(x) > 0$ for all x in $[a, b]$. If the function g is defined by $g(x) = 1/(M-f(x))$, then g is continuous in $[a, b]$, using result (iii) in Example 44, since the denominator is never zero. By Theorem 23 we know that g is bounded, so $1/(M-f(x)) < K$ for x in $[a, b]$, where $K > 0$. But then $M-f(x) > 1/K$, which gives $f(x) < M-(1/K)$. This is a contradiction to the fact that $M = \sup S$, so the original supposition is false.

Example 47. It is essential, in the statement of Theorem 23, that the interval be a closed one. If we consider, for example, the function f given by $f(x) = 1/x$, then in the open interval $(0, 1)$, f is well defined and continuous, but not bounded.

Example 48. A function which is bounded but does not attain its bounds is the function f defined by $f(x) = x-[x]$. If we take the closed interval $\left[\frac{1}{2}, \frac{5}{2}\right]$, say, then the set S of values that $f(x)$ takes is the interval $[0, 1)$, closed at one end and open at the other. So $M = \sup S = 1$, though at no point of the interval does $f(x) = 1$. The supremum is not attained though the infimum is. This function, of course, does not satisfy the condition of Theorem 24 of being continuous.

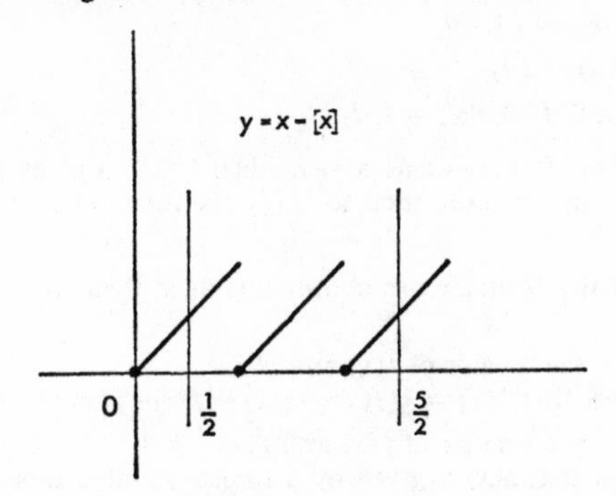

FIGURE 2

Example 49. The same function f given by $f(x) = x-[x]$ is continuous, however, in the interval (1, 2), say, and still fails to attain its bounds; but in this case we are not satisfying the condition of having a closed interval.

EXERCISES

1. Give a definition, following those in the text, of what is meant by $f(x) \to \infty$ as $x \to a$. Give also the form of definition required as $x \to a$ 'from the right' and 'from the left'. Show that $1/(x^2-x-2) \to \infty$ as $x \to 2+$.

2. Draw the graph of $y = f(x)$, where $f(x) = [x]+[-x]$.

3. Find, if it exists, each of the following limits:

(i) $\lim\limits_{x\to 1+} \dfrac{1}{[x]-1}$ (ii) $\lim\limits_{x\to 1-} \dfrac{1}{[x]-1}$

(iii) $\lim\limits_{x\to -1+} \dfrac{1}{|[x]|-1}$ (iv) $\lim\limits_{x\to -1-} \dfrac{1}{|[x]|-1}$

(v) $\lim\limits_{x\to -1+} \dfrac{1}{[|x|]-1}$ (vi) $\lim\limits_{x\to -1-} \dfrac{1}{[|x|]-1}$

4. Prove, from first principles, that $(x-[x])/x \to 0$ as $x \to \infty$.

5. Following the proof of Theorem 15, prove that if $f(x) \to l_1$ and $g(x) \to l_2$ as $x \to a$, then as $x \to a$

(i) $f(x)+g(x) \to l_1+l_2$

(ii) $f(x)g(x) \to l_1 l_2$

(iii) if $l_2 \neq 0$, $f(x)/g(x) \to l_1/l_2$.

6. Prove that if $f(x) \to l$ as $x \to a$, then $|f(x)| \to |l|$ as $x \to a$. Deduce that if f is continuous at a, then so is $|f|$, the function defined by $|f|(x) = |f(x)|$.

7. Prove that if f and g are continuous at a, then so is the function h defined by:

(i) $h(x) =$ the greater of $f(x)$ and $g(x)$.
[Hint: show that $h(x) = \frac{1}{2}(f(x)+g(x)+|f(x)-g(x)|)$.]

(ii) $h(x) =$ the smaller of $f(x)$ and $g(x)$.
[Hint: show that $h(x)$ is given by a similar suitable expression.]

8. Prove that if f is continuous in $[0, 1]$ and $f(0) = f(1)$, then there is a t in $[\frac{1}{2}, 1]$ such that $f(t) = f(t-\frac{1}{2})$.

9. Deduce from Theorems 22 and 24 that if a function is continuous in a closed interval, then it takes every value between its bounds.

CHAPTER FIVE

Differentiable Functions

---

19. Derivatives

The whole of differential calculus develops from one simple idea: a function is said to be *differentiable* if a certain limit exists.

DEFINITION. *The function f is* **differentiable** *at a if*

$$\frac{f(x)-f(a)}{x-a}$$

has a limit as $x \to a$. This limit is denoted by $f'(a)$ and is the **derivative** *of f at a.*

Example 50. If $f(x) = k$ for all x, then (if $x \neq a$), $(f(x)-f(a))/(x-a) = 0$. Thus $f'(a) = 0$.

Example 51. If $f(x) = x$ for all x, then (if $x \neq a$), $(f(x)-f(a))/(x-a) = 1$. Thus $f'(a) = 1$.

Example 52. If $f(x) = x^2$ for all x, then (if $x \neq a$), $(f(x)-f(a))/(x-a) = (x^2-a^2)/(x-a) = x+a \to 2a$ as $x \to a$. Thus $f'(a) = 2a$.

If f is to be differentiable, then in the first place $f(x)$ must be defined for values of x near to a on both sides as well as at a. In fact, we can say very much more than that:

THEOREM 25. *If f is differentiable at a, then f is continuous at a.*

Proof. Differentiability implies that $(f(x)-f(a))/(x-a)$ tends to a limit l as $x \to a$. But $(x-a) \to 0$ as $x \to a$. Therefore (using Exercise 5(ii), Chapter 4),

$$f(x)-f(a) = (x-a) \, . \frac{f(x)-f(a)}{x-a} \to 0.l = 0.$$

Thus $f(x)-f(a) \to 0$, so $f(x) \to f(a)$ as $x \to a$. Hence f is continuous at a.

Example 53. It is quite possible for a function to be continuous at a point without being differentiable. For example, if $f(x) = |x|$ for all x, then f is continuous. For $x > 0$, f is differentiable, and $f'(x) = 1$ for all $x > 0$. For $x < 0$, f is differentiable, and $f'(x) = -1$ for all $x < 0$. At 0, f is continuous but not differentiable.

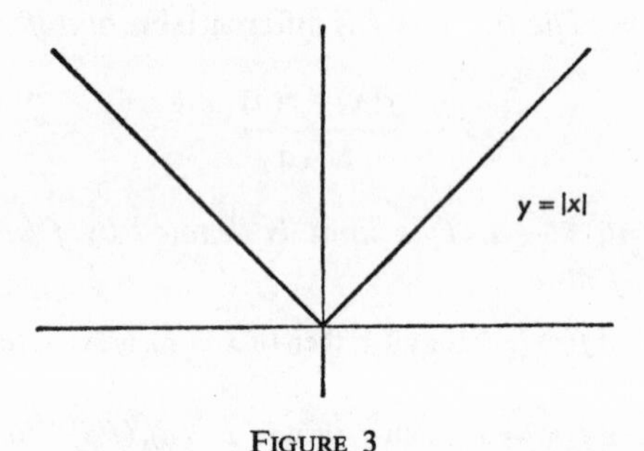

FIGURE 3

If $\lim_{x \to a+} (f(x)-f(a))/(x-a)$ exists, we may call this the **right-derivative** at a, and if $\lim_{x \to a-} (f(x)-f(a))/(x-a)$ exists, this is the **left-derivative**. It is clear that f is differentiable at a point if the left-derivative and right-derivative both exist and are equal.

Example 54. If $f(x) = |x|$ for all x (see Example 53), then at 0 the right-derivative exists and equals $+1$, while the left-derivative exists and equals -1.

THEOREM 26.

(i) *If* $f'(a) > 0$, *there is a neighbourhood of a within which* $f(x) < f(a)$ *when* $x < a$, *and* $f(x) > f(a)$ *when* $x > a$.

(ii) *If* $f'(a) < 0$, *there is a neighbourhood of a within which* $f(x) > f(a)$ *when* $x < a$, *and* $f(x) < f(a)$ *when* $x > a$.

Proof. (i) If $(f(x)-f(a))/(x-a)$ tends to a limit greater than 0 as $x \to a$, there is a neighbourhood of a within which $(f(x)-f(a))/(x-a)$ is positive. Within that neighbourhood, if $x > a$, $x-a$ is positive so $f(x)-f(a)$ is positive, and if $x < a$, $x-a$ is negative so $f(x)-f(a)$ is negative.

(ii) is proved similarly.

In ascertaining the behaviour of a function, it is often useful to consider also the second derivative:

DEFINITION. *If* f *is differentiable at* a *and* f' *is also differentiable at* a, *then the derivative of* f' *at* a *is denoted by* $f''(a)$ *and is called the* **second derivative** *of* f *at* a.

Example 55. If f is defined by $f(x) = x^2$ for all x, then $f'(x) = 2x$ for all x, and $f''(x) = 2$ for all x.

Example 56. If f is defined by $f(x) = |x|$ for all x, then $f'(x) = -1$ for $x < 0, f'(x) = 1$ for $x > 0$. Thus $f''(x)$ is defined for all $x \neq 0$ and $f''(x) = 0$ (for $x \neq 0$).

20. Rolle's theorem

A basic theorem to any further development in differential calculus is the following. Geometıically, it states an 'obvious' result. If two points P and Q, on a reasonably behaved curve $y = f(x)$, are level with each other, i.e. have the same y-co-ordinate, then there is a point on the curve, between P and Q, at which the tangent to the curve is horizontal. We shall, of course, state now what precisely we mean by 'reasonably behaved'.

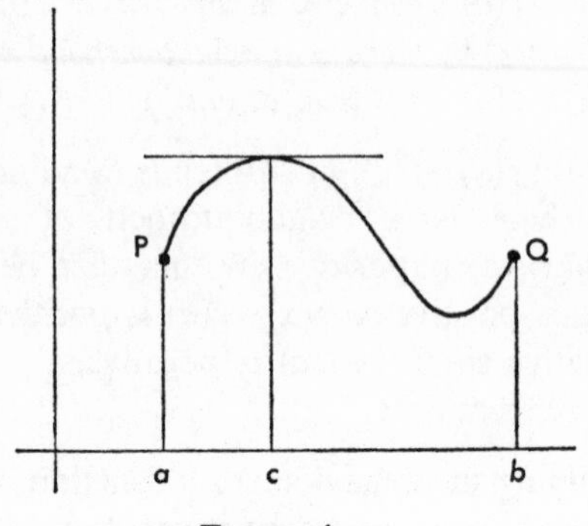

FIGURE 4

THEOREM 27. **(Rolle's theorem).** *If*

(i) f *is continuous in the closed interval* $[a, b]$,

(ii) f *is differentiable in the open interval* (a, b),

(iii) $f(a) = f(b)$,

then there is a number c, *strictly between* a *and* b, *such that* $f'(c) = 0$.

Proof. Since f is continuous in $[a, b]$, we can let $M = \sup S$ and $m = \inf S$, where S is the set of values of $f(x)$ for x in $[a, b]$. If $M = f(a)$ and $m = f(a)$, then $f(x)$ must be constant throughout the interval and $f'(c) = 0$ for any c between a and b. So we may suppose that either $M > f(a)$ or $m < f(a)$. We shall consider the first case. By Theorem 24, let c be the point, strictly between a and b, such that $f(c) = M$.

Suppose, if possible, that $f'(c) > 0$. Then, by Theorem 26, there is a neighbourhood of c within which $f(x) > f(c)$ if $x > c$. This means that there are points, just to the right of c, within $[a, b]$, where $f(x) > M$, a contradiction. Similarly, if $f'(c) < 0$, there are points, just to the left of c, where $f(x) > M$, a contradiction.

Hence $f'(c) = 0$ as required.

Example 57. We could replace (i) and (ii) in Rolle's theorem by insisting, instead, that f is differentiable in the closed interval $[a, b]$. This requirement, however, is unnecessarily strong, and there are functions satisfying (i) and (ii) which are not differentiable in the closed interval. A simple example is f given by $f(x) = +\sqrt{(2x-x^2)}$. This function is continuous in $[0, 2]$, and $f'(x)$ exists between 0 and 2. However, at each end-point, $f'(x)$ does not exist (it becomes infinite). The graph of the function is, in fact, a semi-circle.

21. The mean value theorem

Let us consider now a geometrical picture like the one described for Rolle's theorem, in which the two points P, Q on a 'reasonably behaved' curve are not necessarily level with each other. We may well suspect a comparable result: that there

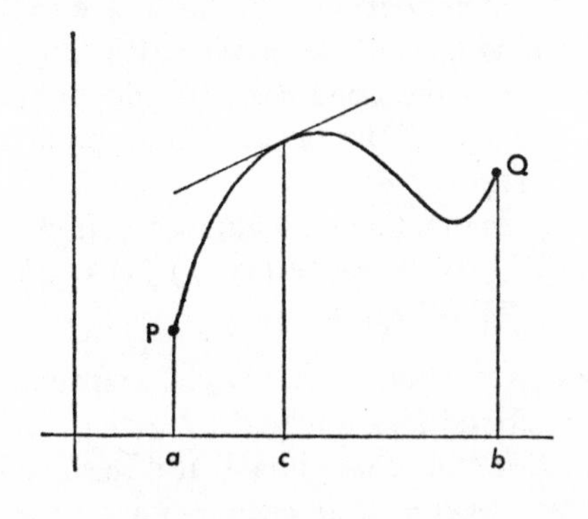

FIGURE 5

is a point on the curve, between P and Q, at which the tangent to the curve is parallel to the straight line PQ. This result can indeed be deduced from Rolle's theorem:

THEOREM 28. **(The mean value theorem).** *If $a < b$ and*
(i) *f is continuous in the closed interval $[a, b]$,*
(ii) *f is differentiable in the open interval (a, b),*
then there is a number c, strictly between a and b, such that

$$f'(c) = \frac{f(b)-f(a)}{b-a}.$$

Proof. Define a new function ϕ by

$$\phi(x) = f(x) - \left(\frac{x-a}{b-a}\right)(f(b)-f(a)).$$

(The idea behind this is to subtract from $f(x)$ an amount proportional to the distance of x from a, so that at a itself nothing is subtracted and at b, $f(b)-f(a)$ is subtracted. In this way we have, closely related to the old curve, a new curve whose end-points are now level with each other.)

Then ϕ satisfies the conditions (i) and (ii) because f does. Moreover, ϕ satisfies the additional hypothesis for Rolle's theorem since $\phi(a) = f(a)$ and $\phi(b) = f(b)-1(f(b)-f(a)) = f(a)$. We may conclude that there is a number c, strictly between a and b, such that $\phi'(c) = 0$. But $\phi'(x) = f'(x)-(f(b)-f(a))/(b-a)$. Thus $f'(c) = (f(b)-f(a))/(b-a)$.

This important result has several immediate applications which can be found in Exercises 7, 8 and 10. One corollary is a rather surprising one: we know that if in some interval $f(x)$ is constant, then $f'(x) = 0$ at each point of the interval. The converse, that if the derivative of a function is identically zero then the function is constant, is not in itself surprising, but it rather unexpectedly needs the mean value theorem to prove it:

THEOREM 29. *If f is continuous in $[a, b]$ and differentiable in (a, b) with $f'(x) = 0$ for all x in (a, b), then $f(x)$ is constant throughout* $[a, b]$.

Proof. For any $x > a$ in $[a, b]$, there is, by the mean value theorem, a number c between a and x such that

$$\frac{f(x)-f(a)}{x-a} = f'(c).$$

But $f'(c) = 0$ and so $f(x)-f(a) = 0$. Consequently $f(x) = f(a)$, and $f(x)$ is equal to this constant value throughout $[a, b]$.

EXERCISES

1. Find from first principles the derivative (stating clearly when it exists) of each of the functions f defined by

   (i) $f(x) = [x]$

   (ii) $f(x) = 1/x \ (x \neq 0)$

   (iii) $f(x) = 1/x^2 \ (x \neq 0)$

   (iv) $f(x) = \sqrt{x} \ (x > 0)$ [Hint: $x-a = (\sqrt{x}-\sqrt{a})(\sqrt{x}+\sqrt{a})$]

   (v) $f(x) = x^n$, where n is a positive integer.
   [Hint: $x^n-a^n = (x-a)(x^{n-1}+x^{n-2}a+\ldots+a^{n-1})$.]

2. Prove that

   (i) if $h(x) = kf(x)$ for all x, then $h'(x) = kf'(x)$

   (ii) if $h(x) = f(x)+g(x)$ for all x, then $h'(x) = f'(x)+g'(x)$

   (iii) if $h(x) = f(x)g(x)$ for all x, then $h'(x) = f(x)g'(x)+f'(x)g(x)$

   (iv) if $h(x) = 1/f(x)$ and $f(x) \neq 0$ for all x, then $h'(x) = -f'(x)/(f(x))^2$.

3. Find constants a, b, c, d so that the function f, defined by $f(x) = -x^2$ if $x < 0$, $f(x) = a+bx+cx^2+dx^3$ if $0 \leq x \leq 1$, $f(x) = x$ if $x > 1$, has a continuous derivative for all x.

4. Verify that the function f, defined by $f(x) = x^2$ if x is rational and $f(x) = 0$ if x is irrational, is differentiable at 0 but not at any other point.

5. We say that f has a **maximum** at a if there is a neighbourhood of a in which $f(x) < f(a)$ except when $x = a$. Similarly, f has a **minimum** at a if there is a neighbourhood of a in which $f(x) > f(a)$ except when $x = a$. Show that if $f'(a)$ exists and f has a maximum or a minimum at a, then $f'(a) = 0$. Verify that $f(x) = x^3$ is an example which shows that the converse is not true.

6. If $f''(x)$ exists for all x in $[0, 2]$ and $f(0) = f(1) = f(2)$, prove that $f''(x) = 0$ for some x in $[0, 2]$.

7. Using the mean value theorem, prove that if f is continuous in $[a, b]$ and differentiable in (a, b), then

(i) if $f'(x) > 0$ throughout (a, b), f is strictly increasing in $[a, b]$ (i.e. if $a \leq x_1 < x_2 \leq b$, then $f(x_1) < f(x_2)$, cf. definition immediately preceding Theorem 36)

(ii) if $f'(x) < 0$ throughout (a, b), f is strictly decreasing in $[a, b]$ (similar definition).

8. Prove that if f'' exists and is continuous in a neighbourhood of a and $f'(a) = 0$, then

(i) if $f''(a) < 0$, f has a maximum at a

(ii) if $f''(a) > 0$, f has a minimum at a.

[Hint: apply Theorem 26 to f' in place of f and use Exercise 7.]

9. Let c be any point in (a, b). Prove that if $f''(x)$ exists and $f''(x) > 0$ for all x in (a, b), then, throughout the interval, the curve $y = f(x)$ lies above the tangent at c. Show also that if $f''(x) < M$ for all x in (a, b), then at any point in the interval, the vertical distance between the curve and the tangent to the curve at c is less than $M(b-a)^2$.

10. Suppose that f and g are continuous in $[a, b]$ and differentiable in (a, b), with $g'(x) \neq 0$ in (a, b). By Rolle's theorem, prove that $g(b)-g(a) \neq 0$. Now, by considering ϕ where

$$\phi(x) = f(x) - \frac{g(x)-g(a)}{g(b)-g(a)}(f(b)-f(a)),$$

prove that there is a number c, strictly between a and b, such that

$$\frac{f'(c)}{g'(c)} = \frac{f(b)-f(a)}{g(b)-g(a)}.$$

Deduce that if $f(a) = g(a) = 0$ then

$$\lim_{x\to a} \frac{f(x)}{g(x)} = \lim_{x\to a} \frac{f'(x)}{g'(x)}$$

if the right-hand side exists.

11. Use the last result of Exercise 10 to find

(i) $\displaystyle\lim_{x\to \frac{1}{2}} \frac{(1-x)^5-x^5}{(1-x)^4-x^4}$

(ii) $\displaystyle\lim_{x\to 1} \frac{2x^4-3x^3+x}{(1-x)^2}$.

CHAPTER SIX

The Riemann Integral

22. Introduction

It is common, at an elementary level, to define integration as the process that is the inverse of differentiation. Thus, given a function f, we look for a function F such that $F'(x) = f(x)$ for all x. There is, in fact, no reason why such a function should exist. Next, if a and b are real numbers with $a < b$, it is shown that $F(b)-F(a)$ gives the value of what is intuitively understood to be the area under the curve $y = f(x)$ between $x = a$ and $x = b$. This approach is unsatisfactory first because it gives no meaning to 'integral' for a function which is not a derivative, and second because it cannot cope with situations where the notion of area under the curve has no obvious intuitive meaning.

Example 58. Elementary methods do not enable you to calculate $\int_0^3 [x]\,dx$. If you look for a function F such that $F'(x) = [x]$ for all x, then F must be continuous, and have gradient 0 between $x = 0$ and $x = 1$, gradient 1 between $x = 1$ and $x = 2$, and gradient 2 between $x = 2$ and $x = 3$. A function satisfying these conditions is given by the following:

$$\begin{aligned} F(x) &= 0, & 0 \le x \le 1, \\ F(x) &= x-1, & 1 \le x \le 2, \\ F(x) &= 2x-3, & 2 \le x \le 3. \end{aligned}$$

But this does not give $F'(x) = [x]$ for *all* x, because $F'(x)$ does not exist at $x = 1$ and $x = 2$.

So this approach does not yield an answer, although you would expect

$\int_a^b [x]\,\mathrm{d}x$ to exist for any a and b; for example you would expect to be able to show that $\int_0^3 [x]\,\mathrm{d}x = 3$.

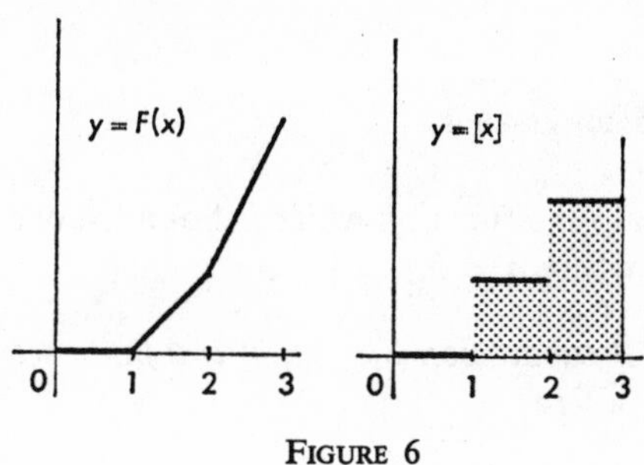

FIGURE 6

Example 59. The method which we shall follow will enable us (see Example 65) to find $\int_1^3 f(x)\,\mathrm{d}x$ where f is given by $f(x) = 1$, for $1 \leq x < 2$ and $2 < x \leq 3$, $f(2) = 3$. This, of course, is not the derivative of any function, so 'anti-differentiation' is not possible.

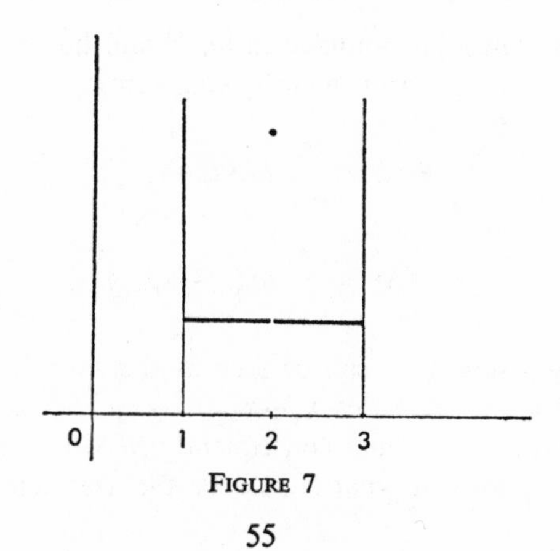

FIGURE 7

So we prefer to define integral in terms of certain summations. This is then taken as the definition of the area under a curve, and it can be related to the process which is the inverse of differentiation in cases where this is meaningful.

23. Upper and lower sums

We suppose that f is a function which is bounded over the closed interval $[a, b]$, where $a < b$.

DEFINITION. *A* **subdivision** Δ *of* $[a, b]$ *is a finite set of numbers* $x_0, x_1, \ldots, x_n$ *where*

$$a = x_0 < x_1 < \ldots < x_n = b.$$

Now let M_r and m_r be the supremum and infimum of the set of values that $f(x)$ takes in the subinterval $[x_{r-1}, x_r]$:

$$M_r = \sup \{f(x) | x_{r-1} \leq x \leq x_r\},$$
$$m_r = \inf \{f(x) | x_{r-1} \leq x \leq x_r\}.$$

These exist since f is bounded in $[a, b]$ and hence in each subinterval $[x_{r-1}, x_r]$. Form the following sums:

$$S(\Delta) = \sum_{r=1}^{n} M_r(x_r - x_{r-1}),$$

$$s(\Delta) = \sum_{r=1}^{n} m_r(x_r - x_{r-1}).$$

The **upper sum** $S(\Delta)$ can be seen as the sum of the areas of n rectangles, on the bases $x_1 - x_0, x_2 - x_1, \ldots, x_n - x_{n-1}$ with heights $M_1, \ldots, M_n$, and consequently $S(\Delta)$ is greater than or equal to our idea of what should be the area under the curve

$y = f(x)$. The **lower sum** $s(\Delta)$ uses rectangles with heights $m_1, \ldots, m_n$, and so is less than or equal to our preconceived idea of the area.

Since f is bounded, there is a number K such that $|f(x)| \le K$ throughout $[a, b]$. Then, using Theorem 6(i) and (ii), we have $|S(\Delta)| \le \sum |M_r||x_r - x_{r-1}| \le \sum K(x_r - x_{r-1}) = K(b-a)$. So $S(\Delta)$ is bounded, above and below. We can therefore take S to be the infimum of the set of all possible upper sums $S(\Delta)$ corresponding to all possible subdivisions Δ:

$$S = \inf \{S(\Delta) | \Delta \text{ is a subdivision of } [a, b]\}.$$

This S is called the **upper Riemann integral** of f over $[a, b]$. Similarly, the **lower Riemann integral** s is equal to the supremum of the set of all possible lower sums:

$$s = \sup \{s(\Delta) | \Delta \text{ is a subdivision of } [a, b]\}.$$

Example 60. If $f(x) = k$ for $a \le x \le b$, then for any subdivision Δ, $M_r = m_r = k$ for all r and $S(\Delta) = s(\Delta) = \sum k(x_r - x_{r-1}) = k(b-a)$. Thus $S = s = k(b-a)$.

Example 61. Let f be the function defined in the interval $[a, b]$ by

$$f(x) = 0, \text{ when } x \text{ is rational,}$$
$$f(x) = 1, \text{ when } x \text{ is irrational.}$$

Within any subinterval $[x_{r-1}, x_r]$ of a subdivision Δ of $[a, b]$, there is always a rational number and an irrational number, so $M_r = 1$ and $m_r = 0$. Thus $S(\Delta) = \sum 1(x_r - x_{r-1}) = b-a$ and $s(\Delta) = \sum 0(x_r - x_{r-1}) = 0$. Thus $S = b-a$ and $s = 0$.

24. Riemann-integrable functions

We shall define a function to be integrable if the upper and lower Riemann integrals are equal, but we must first establish (in Theorem 34, below) that the lower is indeed always less than or equal to the upper.

THEOREM 30. *For any* Δ, $s(\Delta) \leq S(\Delta)$.

Proof. $S(\Delta)-s(\Delta) = \sum (M_r-m_r)(x_r-x_{r-1}) \geq 0$, since every term in the summation is greater than or equal to zero.

THEOREM 31. *If* Δ *is any subdivision of* $[a, b]$ *and* Δ_1 *is the subdivision* Δ *with an additional point, then*

$$s(\Delta) \leq s(\Delta_1) \text{ and } S(\Delta_1) \leq S(\Delta).$$

Proof. If Δ is the subdivision $a = x_0 < x_1 < \ldots < x_n = b$ and y is the additional point, then for some i, $x_{i-1} < y < x_i$ and Δ_1 is the subdivision

$$a = x_0 < x_1 < \ldots < x_{i-1} < y < x_i < \ldots < x_n = b.$$

By definition, $M_i = \sup \{f(x)|x_{i-1} \leq x \leq x_i\}$. To deal with the subdivision with the extra point, we need to introduce $M_i' = \sup \{f(x)|x_{i-1} \leq x \leq y\}$ and $M_i'' = \sup \{f(x)|y \leq x \leq x_i\}$. Then $M_i' \leq M_i$ and $M_i'' \leq M_i$ (see Exercise 1). Consequently,

$$\begin{aligned} S(\Delta_1) &= \sum_{r=1}^{i-1} M_r(x_r-x_{r-1}) + M_i'(y-x_{i-1}) \\ &\qquad + M_i''(x_i-y) + \sum_{r=i+1}^{n} M_r(x_r-x_{r-1}) \\ &\leq \sum_{r=1}^{i-1} M_r(x_r-x_{r-1}) + M_i(y-x_{i-1}) \\ &\qquad + M_i(x_i-y) + \sum_{r=i+1}^{n} M_r(x_r-x_{r-1}) \\ &= \sum_{r=1}^{n} M_r(x_r-x_{r-1}) = S(\Delta). \end{aligned}$$

So $S(\Delta_1) \leq S(\Delta)$. Similarly, with the obvious notation, $m_i \leq m_i'$ and $m_i \leq m_i''$, and $s(\Delta) \leq s(\Delta_1)$.

We may describe this result in words by saying that the introduction of an additional point into a subdivision decreases $S(\Delta)$ and increases $s(\Delta)$. Applying the theorem a finite number of times gives the following:

THEOREM 32. *If* Δ *is any subdivision of* $[a, b]$, *and* Δ_1 *is the subdivision* Δ *with a finite number of additional points, then* $s(\Delta) \leq s(\Delta_1)$ *and* $S(\Delta_1) \leq S(\Delta)$.

Next, we show that any one lower sum is less than or equal to any other upper sum:

THEOREM 33. *If* Δ_1 *and* Δ_2 *are any two subdivisions of* $[a, b]$, *then* $s(\Delta_1) \leq S(\Delta_2)$.

Proof. Let Δ_3 be the subdivision consisting of all the points of Δ_1 and all the points of Δ_2. Then Δ_3 is Δ_1 with a finite number of additional points, so by the preceding theorem $s(\Delta_1) \leq s(\Delta_3)$. However, Δ_3 can also be seen as Δ_2 with a finite number of additional points, so $S(\Delta_3) \leq S(\Delta_2)$. But we established in Theorem 30 that $s(\Delta_3) \leq S(\Delta_3)$. So, combining these three inequalities, $s(\Delta_1) \leq S(\Delta_2)$.

THEOREM 34. *If* s *and* S *are the lower and upper Riemann integrals of* f *over* $[a, b]$, *then* $s \leq S$.

Proof. The previous theorem has established that, if Δ_1 and Δ_2 are any two subdivisions, $s(\Delta_1) \leq S(\Delta_2)$. Consider $s(\Delta_1)$. It is less than or equal to all upper sums $S(\Delta_2)$, where Δ_2 is any subdivision, and is therefore a lower bound of the set of all upper sums. So $s(\Delta_1) \leq S$, since S is the greatest lower bound of the set of all upper sums. But Δ_1 is just any subdivision, so we have shown that S is greater than or equal to all the lower sums. So S is an upper bound of the set of all lower sums, and thus S is greater than or equal to s, the least upper bound of this set, i.e. $s \leq S$.

Now we are in a position to make the definition that we were aiming at:

DEFINITION. *The function f is* **Riemann-integrable** *over the interval* $[a, b]$ *if, in the established notation,* $s = S$. *The common value of* s *and* S *is denoted by*

$$\int_a^b f \, dx.$$

25. Examples

Example 62. From Example 60, we see that if $f(x) = k$ for $a \leq x \leq b$, then f is Riemann-integrable over $[a, b]$ and

$$\int_a^b f \, dx = k(b-a).$$

Example 63. If $a < b$, and f is the function with $f(x) = 0$, when x is rational, and $f(x) = 1$, when x is irrational (see Example 61), then f is not Riemann-integrable.

Example 64. If $f(x) = x$, for all x, then f is Riemann-integrable over $[0, 1]$ (in fact over any finite interval $[a, b]$) and, with a slight abuse of notation,

$$\int_0^1 x \, dx = \tfrac{1}{2}.$$

For, if Δ is the subdivision of $[0, 1]$ with n equal subintervals,

$$0 < 1/n < 2/n < \ldots < (n-1)/n < 1,$$

we have $x_r - x_{r-1} = 1/n$, for all r, $M_r = r/n$ and $m_r = (r-1)/n$. So

$$S(\Delta) = \frac{1}{n}\left(\frac{1}{n} + \frac{2}{n} + \ldots + \frac{n}{n}\right)$$

$$= \frac{1}{n^2}(1+2+\ldots+n)$$

$$= \frac{1}{n^2}\frac{1}{2}n(n+1) = \frac{n(n+1)}{2n^2} = \frac{1}{2} + \frac{1}{2n},$$

and
$$s(\Delta) = \frac{1}{n}\left(0 + \frac{1}{n} + \ldots + \frac{n-1}{n}\right)$$
$$= \frac{1}{n^2}(1+2+\ldots+(n-1))$$
$$= \frac{1}{n^2}\frac{1}{2}(n-1)n = \frac{n(n-1)}{2n^2} = \frac{1}{2} - \frac{1}{2n}.$$

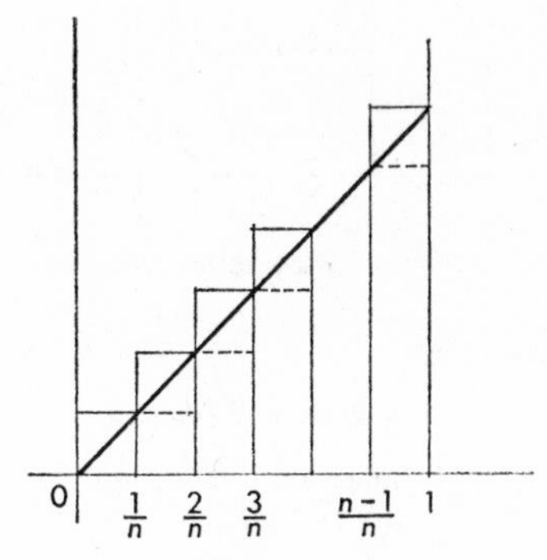

FIGURE 8

By taking n sufficiently large we can make $S(\Delta)$ as near to $\frac{1}{2}$ as we like, so $S \leq \frac{1}{2}$. But, similarly, by taking n sufficiently large, we can make $s(\Delta)$ as near to $\frac{1}{2}$ as we like, so $s \geq \frac{1}{2}$. But since $s \leq S$, we must have $s = S = \frac{1}{2}$.

Example 65. If $f(x)$ is constant throughout an interval, except that at one point $f(x)$ takes some other value, then f is still Riemann-integrable. We prove this for just a particular numerical example, but a proof of the general case could be obtained on similar lines.

Suppose that f is defined on the interval $[1, 3]$ by $f(x) = 1$ for $1 \leq x < 2$ and $2 < x \leq 3$, and $f(2) = 3$ (see Figure 9).

If Δ is any subdivision of $[1, 3]$ then $m_r = 1$ for all r, and consequently $s(\Delta) = \sum_{r=1}^{n} 1(x_r - x_{r-1}) = x_n - x_1 = 3-1 = 2$. So $s = 2$.

We now know that $S \geq 2$, so, in order to show that $S = 2$, we must be able to choose a suitable subdivision Δ which makes $S(\Delta)$ as near to 2 as we like. That is to say, for any given $\epsilon > 0$, we must find a subdivision Δ such that $S(\Delta) < 2+\epsilon$.

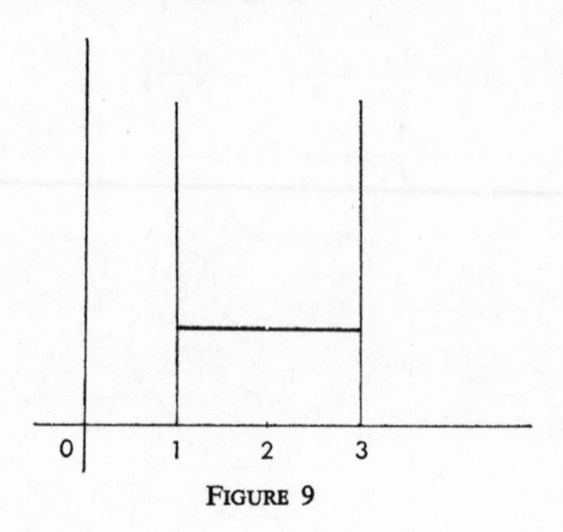

FIGURE 9

Let Δ be the subdivision

$$1 < 2-k < 2+k < 3,$$

where we shall specify k later. Then there are three subintervals, and $M_1 = 1$, $M_2 = 3$, $M_3 = 1$. So

$$\begin{aligned} S(\Delta) &= 1((2-k)-1)+3((2+k)-(2-k))+1(3-(2+k)) \\ &= (1-k)+6k+(1-k) = 2+4k. \end{aligned}$$

Therefore, given $\epsilon > 0$, choose k to be any definite positive number less than $\frac{1}{4}\epsilon$. Then $S(\Delta) = 2+4k < 2+\epsilon$, as required. We have shown that $s = S = 2$.

The value of the integral is thus the same as the value of the integral of the function g where $g(x) = 1$ for $1 \le x \le 3$, showing that the odd point makes no difference. Indeed, you can show that if g is any function which is Riemann-integrable over $[a, b]$ and f is a function whose value differs from that of g at only a finite number of points within the interval, then f is also Riemann-integrable and $\int_a^b f\,dx = \int_a^b g\,dx$.

26. A necessary and sufficient condition

We establish an important necessary and sufficient condition which will be of use in showing that certain functions are Riemann-integrable.

THEOREM 35. *The function f is Riemann-integrable over $[a, b]$ if and only if, given a positive number ϵ, there is a subdivision Δ (depending on ϵ) such that $S(\Delta)-s(\Delta) < \epsilon$.*

Proof. Suppose, first, that for any $\epsilon > 0$, there is such a subdivision Δ. Since $s(\Delta) \leq s \leq S \leq S(\Delta)$, we have $S-s \leq \epsilon$. So $S-s$ is less than any given positive number. Thus $s = S$ and f is Riemann-integrable.

Conversely, suppose $s = S$. Then because s is the supremum of the set of all lower sums, there is a subdivision Δ_1, say, with $s(\Delta_1) > s-\frac{1}{2}\epsilon$. And there is similarly a subdivision Δ_2, say, with $S(\Delta_2) < S+\frac{1}{2}\epsilon$. Let Δ_3 be the subdivision consisting of the points of Δ_1 and the points of Δ_2. Then, by Theorem 32, $s(\Delta_1) \leq s(\Delta_3)$ and $S(\Delta_3) \leq S(\Delta_2)$. So

$$s-\tfrac{1}{2}\epsilon < s(\Delta_3) \leq S(\Delta_3) < S+\tfrac{1}{2}\epsilon.$$

Hence, remembering that $s = S$, we have $S(\Delta_3)-s(\Delta_3) < \epsilon$, and we have found a suitable partition.

27. Monotone functions

We can now show that there is a certain large class of functions which are Riemann-integrable.

DEFINITION. *The function f is* **increasing** *in $[a, b]$ if, whenever $a \leq x_1 < x_2 \leq b$, $f(x_1) \leq f(x_2)$. Similarly, f is* **decreasing** *in $[a, b]$ if, whenever $a \leq x_1 < x_2 \leq b$, $f(x_1) \geq f(x_2)$. A function which is either increasing or decreasing in $[a, b]$ is* **monotone** *in $[a, b]$.*

THEOREM 36. *If f is monotone in $[a, b]$, then f is Riemann-integrable over $[a, b]$.*

Proof. Suppose that f is increasing, say, in $[a, b]$. Let Δ be the subdivision of $[a, b]$ into n equal subintervals, where we shall specify n later. Then $x_r - x_{r-1} = (b-a)/n$. Because f is increasing in each subinterval $[x_{r-1}, x_r]$, $m_r = f(x_{r-1})$ and $M_r = f(x_r)$. So

$$S(\Delta)-s(\Delta) = \sum_{r=1}^{n} M_r(x_r - x_{r-1}) - \sum_{r=1}^{n} m_r(x_r - x_{r-1})$$

$$= \sum_{r=1}^{n} (M_r - m_r)(x_r - x_{r-1})$$

$$= \frac{(b-a)}{n} \sum_{r=1}^{n} (M_r - m_r)$$

$$= \frac{(b-a)}{n} \sum_{r=1}^{n} (f(x_r) - f(x_{r-1}))$$

$$= \frac{(b-a)}{n} (f(b) - f(a)).$$

Now $(b-a)(f(b)-f(a))/n < \epsilon$ if $n > (b-a)(f(b)-f(a))/\epsilon$, so, given $\epsilon > 0$, let Δ be the subdivision of $[a, b]$ into n equal subintervals where $n = [(b-a)(f(b)-f(a))/\epsilon]+1$. Then we have $S(\Delta)-s(\Delta) < \epsilon$, and f is Riemann-integrable by Theorem 35.

Example 66. If $0 \le a < b$ and $a \le x_1 < x_2 \le b$, then we have $x_2^2 - x_1^2 = (x_2 - x_1)(x_2 + x_1) > 0$. Hence the function given by $f(x) = x^2$ is increasing in $[a, b]$ and so it is Riemann-integrable over $[a, b]$.

Example 67. The function in the previous example is a continuous function and we shall show, in Theorem 39, that any continuous function is Riemann-integrable. However, Theorem 36 gives us examples of discontinuous Riemann-integrable functions. For instance, if $f(x) = [x]$, then f is increasing in any finite interval $[a, b]$ and hence Riemann-integrable over $[a, b]$.

28. Uniform continuity

Our next main theorem is to prove that there is another large class of functions, the class of all continuous functions, which are Riemann-integrable. To obtain this, however, we must first investigate continuous functions a little further.

If f is continuous in the closed interval $[a, b]$, we know that it is continuous at each point x_1 in the interval (with right and left continuity at a and b respectively). This means that for any given number $\epsilon > 0$, there is a corresponding δ such that if $|x-x_1| < \delta$ then $|f(x)-f(x_1)| < \epsilon$, where, of course, δ depends on ϵ. But notice, too, that δ depends on x_1, for if we have a different point x_2 in the interval and the same given ϵ, then, possibly, a different smaller δ may be required.

We can establish, however, that there is a δ (depending on ϵ, of course) that will do for any point x_1 in the interval. This notion is called *uniform continuity* and we clarify it below.

DEFINITION. *The function f is said to be* **uniformly continuous** *in the interval $[a, b]$ if, given $\epsilon > 0$, there is a positive number δ such that for any two points x_1, x_2 in $[a, b]$ with $|x_1-x_2| < \delta$, we have $|f(x_1)-f(x_2)| < \epsilon$.*

This is apparently a stronger condition on f than continuity, but we shall show that in a closed interval any continuous function is uniformly continuous.

The following is a convenient term to introduce:

DEFINITION. *The* **variation** *of a bounded function f in an interval $[a, b]$ is equal to*

$$\sup \{f(x)|a \leq x \leq b\} - \inf \{f(x)|a \leq x \leq b\}.$$

THEOREM 37. *If f is continuous in $[a, b]$ then, given $\epsilon > 0$*

there is a subdivision of $[a, b]$ *such that, in each subinterval, the variation of* f *is less than* ϵ.

Proof. Suppose that f is continuous in $[a, b]$ and that, given $\epsilon > 0$, there is *no* subdivision of $[a, b]$ such that in each subinterval the variation of f is less than ϵ. Let c be the midpoint of $[a, b]$ and consider the two intervals $[a, c]$ and $[c, b]$. If each had a subdivision such that in each subinterval the variation of f was less than ϵ, the same would be true of the whole interval, so at least one half does not. Denote the half that does not by $[a_1, b_1]$, i.e. either $a_1 = a$ and $b_1 = c$, or $a_1 = c$ and $b_1 = b$. (If neither half has the stated subdivision, choose, say, the left one.) We may repeat the process with $[a_1, b_1]$ and obtain one half of this interval which does not have a subdivision as described, calling this $[a_2, b_2]$. Continuing, we get an increasing sequence (a_n) and a decreasing sequence (b_n), with $a_n \leq b_n$. By Theorem 16, a_n tends to a limit A and b_n tends to a limit B. But $b_n - a_n = (b-a)/2^n$, so $b_n - a_n \to 0$. Consequently, A is equal to B, so let us write $A = B = \xi$, say.

Now f is continuous at ξ, so there is a δ such that if x is in the interval $(\xi-\delta, \xi+\delta)$, then $|f(x)-f(\xi)| < \frac{1}{2}\epsilon$. If x_1 and x_2 lie in this interval,

$$f(\xi)-\tfrac{1}{2}\epsilon < f(x_1) < f(\xi)+\tfrac{1}{2}\epsilon$$

and

$$f(\xi)-\tfrac{1}{2}\epsilon < f(x_2) < f(\xi)+\tfrac{1}{2}\epsilon,$$

so $|f(x_1)-f(x_2)| < \epsilon$. So the variation of f within $(\xi-\delta, \xi+\delta)$ is less than ϵ. But, because $a_n \to \xi$ and $b_n \to \xi$, there is an N, sufficiently large, such that a_N and b_N lie within $(\xi-\delta, \xi+\delta)$. Hence in $[a_N, b_N]$, the variation of f is less than ϵ, which is a contradiction since, by choice, each $[a_n, b_n]$ has no subdivision with subintervals in which the variation of f is less than ϵ.

THEOREM 38. *If f is continuous in* $[a, b]$, *then f is uniformly continuous in* $[a, b]$.

Proof. Given $\epsilon > 0$, we can, by the previous theorem, find a subdivision of $[a, b]$ so that in each subinterval the variation of f is less than $\frac{1}{2}\epsilon$.

Let δ be the length of the smallest of these subintervals. If x_1 and x_2 are any two points with $|x_1 - x_2| < \delta$, either they both lie in the same subinterval, in which case $|f(x_1) - f(x_2)| < \frac{1}{2}\epsilon$, or they lie in neighbouring subintervals. In the second case, let c be the common end-point of the two subintervals, and we have $|f(x_1) - f(x_2)| \leq |f(x_1) - f(c)| + |f(c) - f(x_2)| < \frac{1}{2}\epsilon + \frac{1}{2}\epsilon = \epsilon$.

29. Integrability of continuous functions

We now use the notion of uniform continuity in the following:

THEOREM 39. *If f is continuous in* $[a, b]$, *then f is Riemann-integrable over* $[a, b]$.

Proof. By Theorem 35, it is sufficient to show that, given $\epsilon > 0$, there is a subdivision Δ such that $S(\Delta) - s(\Delta) < \epsilon$.

Given $\epsilon > 0$, let Δ, for the moment, be any subdivision. Since a continuous function in a closed interval attains its bounds (Theorem 24), for each r, $M_r = f(\xi_r)$ and $m_r = f(\eta_r)$, where ξ_r and η_r are in $[x_{r-1}, x_r]$. Now, using uniform continuity, take the positive number $\epsilon/(b-a)$ and obtain the corresponding δ so that if $|x_1 - x_2| < \delta$, then $|f(x_1) - f(x_2)| < \epsilon/(b-a)$. Now take Δ to be a subdivision whose subintervals have length less than δ. (For example, Δ could be the subdivision of $[a, b]$ into n equal subintervals where $n > (b-a)/\delta$.) Then, since ξ_r and η_r

are in $[x_{r-1}, x_r]$ and $x_r - x_{r-1} < \delta$, we have $|\xi_r - \eta_r| < \delta$ and so $|f(\xi_r) - f(\eta_r)| < \epsilon/(b-a)$. Therefore

$$\begin{aligned} S(\Delta) - s(\Delta) &= \sum_{r=1}^{n} (M_r - m_r)(x_r - x_{r-1}) \\ &= \sum (f(\xi_r) - f(\eta_r))(x_r - x_{r-1}) \\ &< \sum (\epsilon/(b-a))(x_r - x_{r-1}) \\ &= (\epsilon/(b-a)) \sum (x_r - x_{r-1}) \\ &= (\epsilon/(b-a))(b-a) = \epsilon. \end{aligned}$$

So, for this subdivision Δ, $S(\Delta) - s(\Delta) < \epsilon$, as required.

Example 68. Notice that, although Theorems 36 and 39 give us two large classes of functions that are Riemann-integrable, there are Riemann-integrable functions which are neither monotone nor continuous. Such a function was given in Example 65.

30. Properties of the Riemann integral

So far we have only spoken of the integral $\int_a^b f \, dx$, when $a < b$. So, we make the following definitions:

$$\int_a^a f \, dx = 0, \text{ and, if } a > b, \int_a^b f \, dx = -\int_b^a f \, dx.$$

The first thing we shall do next is to consider what happens when the interval $[a, b]$ is split in two, or when two adjacent intervals are combined, and we find that all goes as expected:

THEOREM 40. *If $a < c < b$, then f is Riemann-integrable over $[a, b]$ if and only if f is Riemann-integrable over $[a, c]$ and over $[c, b]$. Moreover, in this situation,*

$$\int_a^c f \, dx + \int_c^b f \, dx = \int_a^b f \, dx.$$

Proof. If f is Riemann-integrable over $[a, b]$, then given

$\epsilon > 0$, there is a subdivision Δ such that $S(\Delta)-s(\Delta) < \epsilon$. Let Δ_1 be the subdivision Δ with the point c added if it does not already belong to A. Then $s(\Delta) \leq s(\Delta_1) \leq S(\Delta_1) \leq S(\Delta)$, so $S(\Delta_1)-s(\Delta_1) < \epsilon$.

Now let Δ_2 be the part of Δ_1 that is in $[a, c]$ and Δ_3 be the part that is in $[c, b]$. Then $S(\Delta_1) = S(\Delta_2)+S(\Delta_3)$ and $s(\Delta_1) = s(\Delta_2)+s(\Delta_3)$. But $s(\Delta_3) \leq S(\Delta_3)$ so $s(\Delta_1)-s(\Delta_2) \leq S(\Delta_1)-S(\Delta_2)$. Therefore $S(\Delta_2)-s(\Delta_2) \leq S(\Delta_1)-s(\Delta_1) < \epsilon$. Because we have found a subdivision Δ_2 of $[a, c]$ such that $S(\Delta_2)-s(\Delta_2) < \epsilon$, f is Riemann-integrable over $[a, c]$. In the same way, we can show that f is Riemann-integrable over $[c, b]$.

Conversely, suppose that f is Riemann-integrable over $[a, c]$ and over $[c, b]$. Let Δ_2, now, be *any* subdivision of $[a, c]$ and Δ_3 be *any* subdivision of $[c, b]$, and let Δ_1 be the subdivision of $[a, b]$ consisting of the points of Δ_2 and Δ_3. Then $S(\Delta_1) = S(\Delta_2)+S(\Delta_3)$ and $s(\Delta_1) = s(\Delta_2)+s(\Delta_3)$.

We *could* argue as follows: because f is Riemann-integrable over $[a, c]$ and over $[c, b]$, we know we can choose Δ_2 such that $S(\Delta_2)-s(\Delta_2) < \frac{1}{2}\epsilon$, and Δ_3 such that $S(\Delta_3)-s(\Delta_3) < \frac{1}{2}\epsilon$. Consequently, $S(\Delta_1)-s(\Delta_1) < \epsilon$ and f is Riemann-integrable over $[a, b]$. But to get the equality between the integrals we must proceed differently.

Let S_1, S_2, S_3 be the upper integrals of f over $[a, b]$, $[a, c]$ and $[c, b]$ respectively. Because $S_2 = \sup\{S(\Delta_2)\}$, given $\epsilon > 0$, we can choose Δ_2 such that $S(\Delta_2) < S_2+\frac{1}{2}\epsilon$, and similarly we can choose Δ_3 such that $S(\Delta_3) < S_3+\frac{1}{2}\epsilon$. Then it follows that $S(\Delta_1) = S(\Delta_2)+S(\Delta_3) < S_2+S_3+\epsilon$. Therefore $S_1 \leq S_2+S_3$.

Similarly, with the lower integrals, we get $s_1 \geq s_2+s_3$. But $s_2 = S_2$ and $s_3 = S_3$, and we know $s_1 \leq S_1$. Hence $s_1 = S_1$, f is Riemann-integrable over $[a, b]$, and

$$\int_a^b f\,dx = \int_a^c f\,dx + \int_c^b f\,dx.$$

As a corollary to this, we have:

THEOREM 41. *If $a \leq c < d \leq b$, and f is Riemann-integrable over $[a, b]$, then f is Riemann-integrable over $[c, d]$.*

Next, with the same interval $[a, b]$ throughout, we see that we get the following expected results, which will constantly be used:

THEOREM 42.

(i) *If f is Riemann-integrable over $[a, b]$, then so is kf, where k is a constant, and*

$$\int_a^b kf \, dx = k \int_a^b f \, dx.$$

(ii) *If f and g are Riemann-integrable over $[a, b]$, then so is $f+g$, and*

$$\int_a^b (f+g) \, dx = \int_a^b f \, dx + \int_a^b g \, dx.$$

(iii) *If f is Riemann-integrable over $[a, b]$, then so is $|f|$.*

(iv) *If f and g are Riemann-integrable over $[a, b]$, then so is fg.*

Proof. (i) The function kf is, of course, the function h where $h(x)$ is equal to k times $f(x)$. This means that we can write this value as $kf(x)$ without confusion.

We shall add f as an upper suffix to the familiar notation used so far when the quantities are those corresponding to the function f, and, in the same way, the appropriate suffix when they correspond to some other function.

First, suppose that $k > 0$. If Δ is any subdivision of $[a, b]$, then $M_r^{kf} = kM_r^f$ and $m_r^{kf} = km_r^f$, so $S^{kf}(\Delta) = kS^f(\Delta)$ and $s^{kf}(\Delta) = ks^f(\Delta)$. Hence $S^{kf} = kS^f$ and $s^{kf} = ks^f$. Thus, if $s^f = S^f = \int_a^b f \, dx$, we have also $s^{kf} = S^{kf} = k \int_a^b f \, dx$.

On the other hand, if $k < 0$, $M_r^{kf} = km_r^f$ and $m_r^{kf} = kM_r^f$.

This implies that $S^{kf} = ks^f$ and $s^{kf} = kS^f$, but, as before, $S^{kf} = s^{kf} = k\int_a^b f\,dx$.

(ii) If Δ is any partition of $[a, b]$, we have $M^{f+g}_r \leq M^f_r + M^g_r$ and so $S^{f+g}(\Delta) \leq S^f(\Delta)+S^g(\Delta)$. It is not difficult to deduce that $S^{f+g} \leq S^f+S^g$. (To do this, it is necessary to choose a subdivision Δ_1 such that $S^f(\Delta_1)$ is less than $S^f+\frac{1}{2}\epsilon$, and a subdivision Δ_2 such that $S^g(\Delta_2)$ is less than $S^g+\frac{1}{2}\epsilon$. If Δ_3 consists of the points of Δ_1 and Δ_2, we find that $S^{f+g}(\Delta_3) \leq S^f+S^g+\epsilon$, and hence obtain the inequality we want.) Similarly, we may show that $s^{f+g} \geq s^f+s^g$. But since $s^f = S^f = \int_a^b f\,dx$ and $s^g = S^g = \int_a^b g\,dx$, and since we know that $s^{f+g} \leq S^{f+g}$, we must have $s^{f+g} = S^{f+g} = \int_a^b f\,dx + \int_a^b g\,dx$.

(iii) It must first be established (we omit the proof) that, if Δ is any subdivision, then, for each r, $M_r^{|f|}-m_r^{|f|} \leq M_r^f-m_r^f$. Therefore, $S^{|f|}(\Delta)-s^{|f|}(\Delta) \leq S^f(\Delta)-s^f(\Delta)$.

If f is Riemann-integrable, then given $\epsilon > 0$, by Theorem 35, there is a subdivision Δ such that $S^f(\Delta)-s^f(\Delta) < \epsilon$. Consequently, with the same subdivision, $S^{|f|}(\Delta)-s^{|f|}(\Delta) < \epsilon$, and, by the same theorem, $|f|$ is Riemann-integrable.

(iv) We first prove that if f is Riemann-integrable, so is f^2. Since f is bounded, there is a positive number K such that $|f(x)| < K$ for all x in $[a, b]$. It can then be shown (proof omitted) that, if Δ is any subdivision, then, for each r, $M_r^{f^2}-m_r^{f^2} \leq 2K(M_r^f-m_r^f)$. Therefore $S^{f^2}(\Delta)-s^{f^2}(\Delta) \leq 2K(S^f(\Delta)-s^f(\Delta))$. If f is Riemann-integrable, given $\epsilon > 0$, there is a subdivision Δ such that $S^f(\Delta)-s^f(\Delta) < \epsilon/2K$. Then, for this subdivision, $S^{f^2}(\Delta)-s^{f^2}(\Delta) \leq 2K(\epsilon/2K) = \epsilon$, and f^2 is Riemann-integrable.

If, now, f and g are Riemann-integrable, notice that $fg = \frac{1}{4}((f+g)^2-(f-g)^2)$, and so fg is Riemann-integrable, from the result just proved together with (i) and (ii).

31. The mean value theorem

THEOREM 43.

(i) *If f is Riemann-integrable over* $[a, b]$, *where* $a < b$, *and* $f(x) \geq 0$ *for* x *in* $[a, b]$, *then* $\int_a^b f \, dx \geq 0$.

(ii) *If* f, g *and* h *are Riemann-integrable over* $[a, b]$, *where* $a < b$, *and* $g(x) \leq f(x) \leq h(x)$ *for all* x *in* $[a, b]$, *then*

$$\int_a^b g \, dx \leq \int_a^b f \, dx \leq \int_a^b h \, dx.$$

(iii) *If f is Riemann-integrable over* $[a, b]$, *where* $a < b$, *and* $m \leq f(x) \leq M$ *for* x *in* $[a, b]$, *then*

$$m(b-a) \leq \int_a^b f \, dx \leq M(b-a).$$

Proof. (i) If Δ is any subdivision, $f(x) \geq 0$ implies that $m_r \geq 0$ for all r. So $s(\Delta) \geq 0$ and $s \geq 0$, and $\int_a^b f \, dx \geq 0$.

(ii) Apply (i) to the functions $f-g$ and $h-f$ and use Theorem 42(i) and (ii).

(iii) Use (ii) with $g(x) = m$ and $h(x) = M$ for all x.

THEOREM 44. **(The first mean value theorem of integral calculus).** *If f is continuous in* $[a, b]$, *then for some value* ξ *in* $[a, b]$

$$\int_a^b f \, dx = f(\xi)(b-a).$$

Proof. Let $\quad m = \inf \{f(x) | a \leq x \leq b\}$

and $\quad M = \sup \{f(x) | a \leq x \leq b\}$.

By Theorem 43(iii),

$$m(b-a) \leq \int_a^b f \, dx \leq M(b-a),$$

i.e.

$$m \leq \frac{1}{b-a} \int_a^b f \, dx \leq M.$$

But m and M are the bounds of the function f, and a continuous function takes every value between its bounds (Exercise 9, Chapter 4). Therefore, for some point ξ in $[a, b]$, we have, as required,

$$f(\xi) = \frac{1}{b-a}\int_a^b f\,\mathrm{d}x.$$

This theorem gets its name from the quantity $\frac{1}{b-a}\int_a^b f\,\mathrm{d}x$, which can reasonably be called the **mean**, or **average, value** of f over the interval $[a, b]$.

32. Integration and differentiation

It remains to establish the relationship between integration and differentiation. This, of course, is of tremendous importance as it connects two extremely powerful ideas in analysis. It also justifies the normal procedure of evaluating integrals by using the reverse process to differentiation when that is possible.

THEOREM 45. *If f is Riemann-integrable over $[a, b]$, and F is defined by*

$$F(t) = \int_a^t f\,\mathrm{d}x, \quad a \le t \le b,$$

then F is a continuous function.

Proof. Let c be any point in $[a, b]$. Since f is bounded, let M be a positive number such that $|f(x)| \le M$ for x in $[a, b]$. Then if t is any point in $[a, b]$,

$$\begin{aligned}|F(c)-F(t)| &= \left|\int_a^c f\,\mathrm{d}x - \int_a^t f\,\mathrm{d}x\right| \\ &= \left|\int_t^c f\,\mathrm{d}x\right|, \quad \text{by Theorem 40,}\end{aligned}$$

$$\leq M|c-t|, \quad \text{by Theorem 43(iii).}$$

So, given $\epsilon > 0$, take δ to be less than ϵ/M and if $|c-t| < \delta$, we have $|F(c)-F(t)| < M\delta < \epsilon$. This means that F is continuous at c.

Now we make a stronger assumption, that f is continuous. We are then able to deduce the fact that F is differentiable and also that its derivative is just f.

THEOREM 46. *If f is continuous in $[a, b]$ and F is defined by*

$$F(t) = \int_a^t f\,dx, \quad a \leq t \leq b,$$

then F is differentiable and $F'(t) = f(t)$ for all t in $[a, b]$.

Proof. Let c be a point in $[a, b]$. We want to show that $(F(c)-F(t))/(c-t)$ tends to a limit as $t \to c$, and that this limit is $f(c)$. Consider

$$\begin{aligned}\frac{F(c)-F(t)}{c-t} &= \frac{1}{c-t}\left\{\int_a^c f\,dx - \int_a^t f\,dx\right\}\\ &= \frac{1}{c-t}\int_t^c f\,dx\\ &= f(\xi),\end{aligned}$$

where ξ is some point between c and t, by Theorem 44.

Therefore

$$\left|\frac{F(c)-F(t)}{c-t} - f(c)\right| = |f(\xi)-f(c)|.$$

Now, because f is continuous at c, given any $\epsilon > 0$, we can find δ such that if $c-\delta < \xi < c+\delta$ then $|f(\xi)-f(c)| < \epsilon$. So, given $\epsilon > 0$, take this δ and take $c-\delta < t < c+\delta$. We shall also have $c-\delta < \xi < c+\delta$, because ξ lies between c and t, and hence

$$\left|\frac{F(c)-F(t)}{c-t} - f(c)\right| < \epsilon.$$

Therefore, $$\lim_{t \to c} \frac{F(c)-F(t)}{c-t} = f(c).$$

Finally, we show that if a function can be found whose derivative is equal to f, then this can be used to evaluate the integral of f.

THEOREM 47. *If f is continuous in $[a, b]$ and ϕ is a function such that $\phi'(x) = f(x)$ for all x in $[a, b]$, then*

$$\int_a^b f \, \mathrm{d}x = \phi(b)-\phi(a).$$

Proof. We have seen, in the previous theorem, that if

$$F(t) = \int_a^t f \, \mathrm{d}x,$$

then $F'(x) = f(x)$ for all x in $[a, b]$. It is given that $\phi'(x) = f(x)$ for all x in $[a, b]$, so $F'(x)-\phi'(x) = 0$. Hence, by Theorem 29, $F(x)-\phi(x) =$ constant. Therefore, $F(b)-\phi(b) = F(a)-\phi(a)$, and consequently

$$\phi(b)-\phi(a) = F(b)-F(a) = \int_a^b f \, \mathrm{d}x.$$

Example 69. The last theorem is the one that justifies the normal procedure for evaluating, say,

$$\int_a^b x \, \mathrm{d}x.$$

If $\phi(x) = \frac{1}{2}x^2$, then $\phi'(x) = x$. Hence

$$\int_a^b x \, \mathrm{d}x = \tfrac{1}{2}b^2 - \tfrac{1}{2}a^2.$$

EXERCISES

1. Show that, if $a \leq c < d \leq b$,

$$\sup \{f(x)|c \leq x \leq d\} \leq \sup \{f(x)|a \leq x \leq b\}$$
$$\inf \{f(x)|c \leq x \leq d\} \geq \inf \{f(x)|a \leq x \leq b\}$$

(used in the proof of Theorem 31).

2. If Δ_n is the subdivision of $[0, 1]$ into n equal subintervals, find $s(\Delta_n)$ and $S(\Delta_n)$ for the function f defined by $f(x) = x^2$. Deduce that

$$\int_0^1 x^2 \, dx$$

exists, and give its value.

[Hint: $\sum_{r=1}^{n} r^2 = \frac{1}{6}n(n+1)(2n+1)$.]

3. If Δ is the subdivision $1 < 2-k < 2+k < 3$ of the interval $[1, 3]$, find $s(\Delta)$ and $S(\Delta)$ for the function f defined by $f(x) = 1$ if $1 \leq x < 2$, $f(x) = 2$ if $2 \leq x \leq 3$. Hence show that

$$\int_1^3 [x] \, dx$$

exists, and give its value.

4. Show that

$$\int_a^b x \, dx$$

exists, and find its value, following the method of Example 64.

5. If $0 < a < b$, let Δ be the following subdivision of $[a, b]$:

$$a < ar < ar^2 < \ldots < ar^{n-1} < ar^n = b.$$

Show that for the function f defined by $f(x) = x$, $s(\Delta) = (b^2-a^2)/(r+1)$ and $S(\Delta) = r(b^2-a^2)/(r+1)$. Show that by taking n sufficiently large, r can be made as near to 1 as we please, and hence obtain the value of

$$\int_a^b x \, dx$$

by a different method from Exercise 4.

6. Give an example to show that $|f|$ Riemann-integrable over $[a, b]$ does not in general imply f Riemann-integrable over $[a, b]$.

7. Supply the detailed proofs of the following results used in the proof of Theorem 42, where sup and inf are taken over some fixed interval:

(i) $\sup\{f(x)+g(x)\} \leq \sup\{f(x)\}+\sup\{g(x)\}$,
$\inf\{f(x)+g(x)\} \geq \inf\{f(x)\}+\inf\{g(x)\}$;

(ii) $\sup\{|f(x)|\}-\inf\{|f(x)|\} \leq \sup\{f(x)\}-\inf\{f(x)\}$;

(iii) $\sup\{f^2(x)\}-\inf\{f^2(x)\} \leq 2K(\sup\{f(x)\}-\inf\{f(x)\})$,

where $|f(x)| < K$.

8. Prove that, if f is Riemann-integrable over $[a, b]$ and ϕ is a function such that $\phi'(x) = f(x)$ for all x in $[a, b]$, then

$$\int_a^b f\,\mathrm{d}x = \phi(b)-\phi(a).$$

[Notice that this is Theorem 47, except that we have a weaker hypothesis: f is not necessarily continuous. Hint: given a subdivision Δ, use the mean value theorem (of differential calculus, Theorem 28) to show that

$$\frac{\phi(x_r)-\phi(x_{r-1})}{x_r-x_{r-1}} = \phi'(y_r) = f(y_r),$$

where y_r is in (x_{r-1}, x_r). Then consider the sum $\sum f(y_r)(x_r-x_{r-1})$ which lies between $s(\Delta)$ and $S(\Delta)$.]

Answers to the Exercises

Chapter One

8. (i) 2, $\sqrt{2}$ (ii) 1, $\frac{1}{2}$ (iii) -1, none (iv) $\sqrt{2}$, $-\sqrt{2}$ (v) -1, -2.

Chapter Two

3. E.g. $a_n = (-1)^n n/(n+1)$.
6. (i) has the limit 1 (ii) oscillates finitely (iii) oscillates infinitely.
7. $A = 1$.
8. E.g. (i) $(a_n) = 1, 2, 1, 4, 1, 8, \ldots$ $(b_n) = -1, -2, -1, -4, -1, -8, \ldots$
   (ii) $(a_n) = 1, 2, 1, 4, 1, 8, \ldots$ $(b_n) = 0, -2, 0, -4, 0, -8, \ldots$
   (iii) $(a_n) = 1, 2, 1, 4, 1, 8, \ldots$ $(b_n) = 1, 2, 1, 4, 1, 8, \ldots$.
9. (c_n) tends to a limit
   (e.g. $(a_n) = 0, 1, 0, 1, 0, 1, \ldots$ $(b_n) = 1, 0, 1, 0, 1, 0, \ldots$)
   or (c_n) oscillates finitely
   (e.g. $(a_n) = 0, 1, 0, 1, 0, 1, \ldots$ $(b_n) = 0, 1, 0, 1, 0, 1, \ldots$).
10. E.g. $a_n = (-1)^n/n$.
13. E.g. $a_n = (n+1)/n$.

Chapter Three

5. $C, D, C, C, C, D, C, D, C$ (C = convergent, D = divergent).
8. Absolutely convergent, divergent, conditionally convergent, divergent.
9. E.g. $1+1+1+1+\ldots$, $-1-1-1-1-\ldots$, $1-1+1-1+1-\ldots$, $1-1+2-2+3-3+\ldots$.
10. Convergent for $-\frac{9}{2} < x < \frac{9}{2}$, $x = 0$, all x, $-3 \leq x < 3$.

ANSWERS TO THE EXERCISES

Chapter Four

2. $f(x) = 0$ for integer values of x, and $f(x) = -1$ for all other values of x.
3. (i) None (ii) -1 (iii) none (iv) 1 (v) -1 (vi) none.

Chapter Five

3. $a = 0$, $b = 0$, $c = 2$, $d = -1$.

11. $\frac{5}{8}$, 3.

Chapter Six

2. $(n-1)n(2n-1)/6n^3$, $n(n+1)(2n+1)/6n^3$, $\frac{1}{3}$.
3. $3-k$, $3+k$, 3.
4. $\frac{1}{2}(b^2-a^2)$.
6. E.g. $f(x) = -1$ when x is rational, $f(x) = 1$ when x is irrational.

Index

INDEX